Springer Geography

The Springer Geography series seeks to publish a broad portfolio of scientific books, aiming at researchers, students, and everyone interested in geographical research. The series includes peer-reviewed monographs, edited volumes, textbooks, and conference proceedings. It covers the entire research area of geography including, but not limited to, Economic Geography, Physical Geography, Quantitative Geography, and Regional/Urban Planning.

More information about this series at http://www.springer.com/series/10180

Shekh Moinuddin

Mediascape and The State

A Geographical Interpretation of Image Politics in Uttar Pradesh, India

Springer

Shekh Moinuddin
Department of Geography
Kalindi College
New Delhi
India

ISSN 2194-315X ISSN 2194-3168 (electronic)
Springer Geography
ISBN 978-3-319-51931-9 ISBN 978-3-319-51932-6 (eBook)
DOI 10.1007/978-3-319-51932-6

Library of Congress Control Number: 2017930269

Printed on acid-free paper

This Springer imprint is published by Springer Nature
The registered company is Springer International Publishing AG
The registered company address is: Gewerbestrasse 11, 6330 Cham, Switzerland

Dedicated to Dr. Aparajita De, Amma and Abba

Preface

This study looks into spatio-temporal aspects of mediascapes and the state with respect to UP Assembly elections 2012 as a case study. The study considered media as space and how and what ways the real space has been portrayed across virtual space (media space) as well as the dichotomy national and regional are as the state. The spatial dimensions were discussed under different chapters with the help of image(s) provided by mediascapes and the state for political reason. The publication is derived from Ph.D. thesis.

The study starts with an introduction to discuss both mediascape and the state—how and at what extent they became phenomenal for political construction in terms of image politics on the one side while on the other it had been supported by conceptual framework and review of literature to understand the image politics. This study draw a bigger picture of political discourse wherein both mediascape and the state played a catalytic role during UP Assembly elections 2012 to know how both mediascape and the state played a phenomenal role to shape and reshape the political discourses in the state.

In second chapter, *Mapping the Mediascapes in India*, the study of "mediascapes" is highlighted in both public and private contexts on how they shaped the business as well as socio-cultural movement in the country. The media landscape includes both old and new media, wherein every segment of media, viz. newspaper, radio, TV and now social media has their own nature and characteristics to serve the people. The role of media increased over the years and is quite evident in everyday activities, and everyone supposes to consume little bit of media every day in different proportions. This chapter also examines the dimensions of political economy of media—how media diverted to serve from public interests to private interests in the country.

Third chapter, *From Regional to National: The Making of UP as the "Heartland" of Indian Politics and Polity*, argues how UP enjoys a privileged political position in Indian politics and media. It also examines how both caste and communal politics have shaped the identity of the state over the years. How Dalit politics, communal violence and electoral spaces had shaped the state as a heartland as well as political consequences of maximum number of parliamentarians or prime ministerial candidates or electoral spaces in mundane representation and interpretations.

Fourth chapter, *Locating Mediated "Heartland" in Indian Politics and Polity*, discusses three issues that appeared across the mediascapes as part of eight issues considered in the study to discuss the image politics of the state. These three issues have own political consequences and have been considered as a political catalyst prior to the election in the state.

Fifth chapter, *Locating Communal Vote Bank Politics in Mediated "Heartland"*, argues how media communalized an issue of reservation for minority appeared across the mediascapes, prior to the election, but to what extent both mediascapes and the state misrepresented and misinterpreted the issue. The myriad political interpretations were sketched including imprints of communal shadow as well.

In sixth chapter, *From National to Regional: Development and Governance as the New Avatars of UP Heartland Politics,* development and governance agenda are discussed along with contentious politics. This chapter deals with four issues which are considered as part of national politics to surround the local government.

The last chapter is about conclusions and findings to draw the inferences together about how media and the state portrayed the UP Assembly elections 2012 and helped politicians to politicize the issues. Media shaped the image of UP as heartland through news that appeared during the election campaign. However, media was successful in propagating the differences between the regional and national politics amongst people.

New Delhi, India Shekh Moinuddin

Acknowledgements

I am thankful to Aparajita De, for her valuable comments, critique and suggestions that shape the works in best possible manner. I would also like to thank R.B. Singh, Anu Kapur, S.C. Rai and Anjan Sen for timely suggestions. I am grateful to University of Delhi and UGC for giving me financial supports to complete the works. Hence, I am extending my sincere thanks to library staffs where I was a frequent visitor—Ratan Tata Library (RTL) and Central Reference Library (CRL) of University of Delhi, Centre for The Study of Developing Societies (CSDS). I am also thankful to Sunil, Tanu, Arvind, Ashar, Sunni, Namita, Rishika and Versha who made his/her valuable contributions and moreover, thankful to both Suhkpreet Kahlon and Rajive De for valuable suggestions and critical comments during final compilation of the draft. Finally, I am extending my strong gratitude to my family members my mother, sisters and brother who constantly supported me in all possible ways.

I am thankful to Muthu Rajan and Michael Leuchner of Springer and the anonymous peer reviewer for suggestions and comments, which improved the book.

Shekh Moinuddin

Contents

Abbreviations

2G	Second Generation
3G	Third Generation
4G	Fourth Generation
AICC	All India Congress Committee
AIR	All India Radio
BJP	Bhartya Janta Party
BSP	Bahujan Samaj Party
CBI	Central Bureau of Investigation
CD	Compact Disc
CNNIBN	Cable News Network Indian Broadcasting Network
DD	Doordarshan
DTH	Direct Television in House
ECI	Election Commission of India
FDI	Foreign Direct Investment
IAMAI	Internet and Mobile Association of India
INC	Indian National Congress
JSY	Janani Suraksha Yojana
MLA	Member of Legislative Assembly
MNREGA	Mahatma Gandhi National Rural Employment Guarantee Programme
NDA	National Democratic Alliance
NDTV	New Delhi Television
NRHM	National Rural Health Mission
OBC	Other Backward Caste
PAG	Principal Auditor General
RLD	Rastriya Lok Dal
SJP	Samajwadi Janta Party
SP	Samajwadi Party
TOI	The Times of India

TRPs	Television Rating Point
TV	Television
UP	Uttar Pradesh
UPA	United Progressive Alliance

List of Figures

List of Images

List of Tables

List of Maps

Chapter 1
Introduction: Mediascape and the State

The image is never a simple reality.

(Ranciere Jacques 2007: 6)

Abstract Mediascape and the state portrayed spatial imageries in real and imaginary shapes to forward arguments to fix a specific political discourse in the spatiality. However, both are dependent on each other or in other word, both are supposed to make interaction in mundane shapes including image politics. During election time, both (mediascape and state) are more phenomenal and catalytic for dissemination of information through texts, visuals and audio. The study considered UP Assembly election 2012 as a case study to code and decode the representational and interpretational politics around image(s) during election campaign.

Keywords Media · State · Spatial · Content analysis · Semiotic analysis

Media is a virtual space where "real" and "imaginary" constructions continually take place and create geographical knowledge that compliments or contests each other simultaneously. Media portrays and represents the image in geographical, political, economic and cultural contexts. It is true that we live in a "mediated" society where many of our ideas about space/place usually come through the virtual space. Media has slowly yet surely, changed the perception of an individual about the geographical spaces/places that we live in. The visual spaces in media have had a great influence on the thinking processes of individual/groups in and about spaces/places. The content of media includes news, entertainment, serials, advertisements and infotainments, etc. that creates experiences and knowledge of space which was hitherto unknown to us and it creates spatial imageries. These spatial imageries might vary in nature and characteristics but are popularized by media.

S. Moinuddin, *Mediascape and The State*, Springer Geography,
DOI 10.1007/978-3-319-51932-6_1

Knut Lundby discusses the overwhelming presence of media in our lives and infers, "The media are everywhere and all-embracing."[1] Nowadays media has become an inevitable phenomena and everybody consumes it irrespective of gender, caste, class, region, religion and tribe. As Knut Lundby points out, "Everyday life and the operations of society depend on mediated communication. Who could manage without a cell phone, e-mail, favorite social networking site, or whatever means of communication one chooses to stay connected?"[2] Whatever the medium is, the presence of media is inevitable in our social life in various capacities. Gillian Rose argues, "with the exception of anthropology, geography is unique in social sciences in the way it has relied and continues to rely on certain kinds of visualities and visual images to construct its knowledge."[3] The visuals shown by media are about analyzing spaces/places in a critical yet contested manner.

Visualities in geography are best articulated and represented through maps which explain space and spatial distributions in a systematic manner. Thornes (2004) in his work points out that the "visual" is the central essence of geography and in contemporary academics; these characteristics give geography an opportunity to take a leading role in exploiting the power of the "visual" in both teaching and research.[4] This visual turn has taken place across most of the disciplines[5] and is no way exclusive to geography.[6] The visuals or images come from different sources wherein media is one of the sources. Media is considered as one of the major sources of spatial imageries that depict spatial phenomena through various contents and programmes. For instance, prior to 2004, the word "tsunami" was not relevant to the Indian context, neither was it common geographical knowledge that the everyday man had. But with the media coverage of the 2004 tsunami, it created geographical knowledge and experience of the tsunami for the common man. At the same time, the visuals of the tsunami created public awareness about its destructiveness which in turn led to state action, adoption of measures and policies for coastal preservation and conservation. Similarly, during the UP Assembly elections 2012, media and the state became one of the most important sources of creating knowledge and making people aware about political discourses and political development in the state. Media coverage created political knowledge and the ground reality for its audience in terms of news that was broadcast and printed.

According to Joe Painter, the state in an everyday sense functions "as an organization with de facto responsibilities for the government and administration of a territory."[7] He reads the modern state as "institutional ensembles", wherein different institutions are supposed to work independently without any superimposition

[1]Lundby (2009).

[2]Ibid.: 1.

[3]Rose (2003).

[4]Thornes (2004).

[5]Mitchell (1992).

[6]Thornes (2004).

[7]Painter (1995, p. 32).

on each other. Painter elaborates, "The sovereign state, we posited, is not the basis for, but effect of, discourse of sovereignty, security, and identity."[8] The state is an agglomeration of different institutions and each institution supposedly works independently without any overlaps and superimpositions to support the state in its day-to-day activities. In democratic set-up, executive, legislative, judiciary and media considered pillar to sustain the ethical, moral and rational interpretation and representation of space/place. However, the role of media expands over period with the expansion of new (social) media.

Kittler (1999) argues that "media determine our situation"[9] while McLuhan (1964) in his seminal work *Understanding Media: The Extensions of Man*, states that "the medium is the message,"[10] shifting our attention away from media context/texts and focusing on technology, which he believed shaped human perceptions. McLuhan (2006) argues that media is a medium to express spatiality in terms of social relations in a particular "content."[11] de Certeau (2009) work on everyday consumption has been widely explored by media analysts as "practice." In our day-to-day practices, media, whether print or electronic, is constantly creating "myths" and "realities" of spaces and geographies in and around us and in different contexts. Media enhances the reach of these visual narratives through their "content" and "circulation."[12] The expansion of media in our everyday life worlds brings not only a wider range of spatial knowledge but the spatial politics of representation that is inherent in what is being displayed and concealed by the media about these spaces.

The representation or narratives around image(s) and texts would have to promote ideological or hegemonic tradition. Durham and Kellner (2001, reprint 2005) argues that "all cultural representations are political is one of major themes of media and cultural theory of the past decades. In the 1960, feminists, African-American, Latino, gay and lesbian, and disparate oppositional movements attacked the stereotypes and biased images of the groups."[13] The politics was the reason (whether positive side or negative side) behind the representations of such image(s). The basic theme of image(s) production was to cater the political needs at best.

The interpretations of image(s) are analyzed in various ways with the help of contents and some other methods to decipher the coding and decoding of portrayals. This study is used UP Assembly election 2012 as image(s) to understand the politics of representation in spatio-politico contexts. However, cultural, ideological and theoretical framework is required to understand the functional style of these institutions (media and the state) for production of such image(s).

---

[8]Kuus and Agnew (2008, pp. 95–106).

[9]Kittler (1999).

[10]McLuhan (1964).

[11]McLuhan (2006).

[12]de Certeau (2009, pp. 76–88).

[13]Durham and Kellner (2001, p. 24).

1.1 Negotiating Media(Scapes)

Media is everywhere. It has been a part of society in various ways. For example, during medieval times, beating the drum was an important source of dissemination of information while in present times; social media has become the fastest disseminator of information. In the past; salons, coffee houses and clubs were the places where people would gather to collect news (Habermas 1989, 2010). By the eighteenth century, the practice of media was not quite public in nature and communication and sharing information amongst people was limited mostly at the personal level. The practices of media and its growth became systematic after the eighteenth century with the emergence of the bourgeois "public sphere", which was controlled by the feudal classes and to some extent the church. The industrial revolution paved the way for democracy and media was shaped by the elite classes of society for their own purpose (during the eighteenth century). Couldry (2012) argues three dimensions of media.[14]

1.1.1 Media as Practice

Nick Couldry infers four basic advantages for media as practice. First as *regularity*, that means "*regularity of action*". He says:

> Sociology itself is interested in regularities, not chance or incidental occurrences. Media sociology is concerned with the specific regularities in our action related to media and the regularities of context and resource that make certain type of media-related actions possible or impossible, likely or unlikely.[15]

Until and unless there are some levels of regularity and order, it is improbable to act at any level. The second type is "*social*". Couldry used Wittgenstein's *philosophy of language* to make use of practice and the way in which language helps us understand the world.

> Behind the recent turn to "practice" in social theory lies an interest in Wittgenstein's later philosophy of language. This involved a key move to understanding language as action in the world, by contrast with an older view of language as the expression of meaning that must somehow "correspond" the world…Practices are not bundles of individual idiosyncrasies; they are social constructions that carry with them a whole world of capacities, constraints and power.[16]

The expression of language is considered the main attribute of media and it envisages the social constructions at the spatial order where fragmentation and unity can easily be accessed to know the social space. The third type is human "*needs*" or

---

[14]Couldry (2012).

[15]Ibid.: 33–4.

[16]Ibid.: 33–4.

the ways in which media becomes instrumental when coordination, interaction, community, trust and freedom shape the media practices. The fourth type is normative, the way in which we should live with media and further Couldry argues that "so best starting point for an inquiry into how we should live with media is to think about media as practice."[17]

On these grounds, Couldry examines "media as practice" and substantiates the social and cultural facts that transform the social space and people who use media as practice in myriad forms. In India, media inevitably becomes practice for people who use media according to their suitability. While K.M. Shrivastava has examined the way in which media has developed over a period of time and how it has constantly expanded its horizons with new aims and objectives, he also argued that during the course of the twentieth century, mankind has travelled from the "Gutenburg Galaxy" to the "Global Village" ushering in information technology.[18] But in India, the expansion of media has taken place at an even faster pace.

Apart from the existing media sources, there are other kinds of media sources known as "new media" and "social media" across the world. In India, social media established its roots at the behest of mainstream media. The role of social media has been witnessed during the "Jasmine revolution" in Morocco, Egypt and the "Lokpal movement" in India. Lyn Gorman and David Mclean argue, "Ongoing, rapid technology change has produced a second wave of new media that may have the potential to change quite radically the relationship between the 'audiences' or 'consumers' and media."[19] So, media as practice was established around us or was supposed to become a medium to convey the spatial messages at various orders.

1.1.2 *Media as Medium*

A medium is the only communicating source between two known and unknown objects. Media is one of the mediums and it exists in a physical state across the globe. Being a medium like other mediums, media carries information or content. Media as medium emerged during the eighteenth century and was established across the world. The present research is aimed at examining the present context of media as medium. How do media interact with people? McLuhan identified media as being an extension of "electric technology", which is enhanced by "automation" and "information retrieval" processing in terms of learning and knowing. McLuhan in his media dictum, "The medium is the message", tries to exalt the knowledge-based content, wherein the content is supposed to define the use of the medium. As he says:

---

[17]Ibid.: 34–5.

[18]Shrivastava (1998).

[19]Gorman and Mclean (2009).

> The instance of the electric light may prove illuminating in this connection. The electric light is pure information. It is a medium without a message, as it were, unless it is used to spell out some verbal ad or name. This fact, characteristics of all media, means that the 'content' of any medium is always another medium. The content of writing is speech, just as the written word is the content of print, and print is the content of telegraph. If it is asked, 'What is the content of speech?' It is necessary to say, 'It is an actual process of thought, which is in itself nonverbal.' An abstract painting represents direct manifestation of creative thought process as they might appear in computer designs.[20]

The content defines the boundary of the medium; and since media is a part of technology that carries purposive information, the information is sliced into various parts. The content of media is the "message" that it carries through various programmes (soap opera, news, infotainment, entertainment, advertisements, religion, sports, etc.). The medium is known by content and content is known by "representation," wherein the representation is the data source of content. The content is situated in media as medium across the globe and through representation it has established media as a medium having "messages" by way of "content" and "representation". Thus, both content and representation stimulate the medium (image) irrespective of shape and size to adhere customized image or in other words, mediated image which is not less than ritual image.

1.1.3 Media as Ritual

Media as ritual has been gradually noticed by scholars at different fronts and they have substantiated their findings through various case studies. Once again, Couldry's concepts are relevant here, about how media functions as ritual through a particular set of programming—especially the reality show. He says, "Media rituals, put simply, are social forms that naturalize media's consistent will-to-power, that is, media's claim to offer privileged access to a common reality to which we must pay attention."[21] He further argues:

> The concept of media rituals starts out from an appreciation of the role of symbols in social ordering. But it takes this in a direction quite opposed to traditional functionalism, foregrounding not the consensual nature of 'the symbolic' but the inherently contested nature of symbolic power.[22]

"Rituals are enactments of power through form."[23] The form may have a varied range of programmes which includes reality shows, news, soap operas,

[20]McLuhan [1964(2001)].

[21]Couldry (2012, p. 66).

[22]Ibid.: 66.

[23]Ibid.: 66.

advertisements, etc. Media in the news business has a different set of rules apart from those segments of media which are running their businesses in other fields (sports, serials, movies, religious, infotainment, etc.) excluding news. This news consists of information about the space which is supposed to make people aware about what happened at whose behest. Such news may be contested to some extent. Symbols (image) which consolidate identity or an ideology are considered the best way to popularize the news/issues.

Nick Couldry further notes, "Media rituals are formalized patterned actions relating to media that enact a particular way of organizing the world."[24] During the UP Assembly election 2012, the media developed a certain order to put news into categories which were more suitable for attracting the audience and politicians alike. Indian media runs a particular set of news in such a manner that all political fulcrums are positioned around those issues.

Media rituals push people to consume issues provided by the media, which have the least chances of discarding the view of the news. The issues that are picked reflect the territory and are relevant in the contemporary context. The media rituals followed are quite contested and symbolic and the discussions that are held over issues put the issue in the public domain in different contexts. As Couldry points out, "How can media institutions command enough attention and legitimacy to be the focus of media rituals anymore?"[25]

Similarly, print media also publishes some sort of news to give a regular story to their readers. For example, *The Times of India* came up with a special page (placed in between the pages), titled "Dance of Democracy", which was firmly dedicated to the political news, particularly centred on the states where the assembly elections were being held. It started a special page to describe the election-based news. This was perhaps to cater to the fact that during election season, people wish to know the political news or perhaps the publishers wish to provide such news to their readers, as news is a powerful symbol in terms of photographs (images) and makes the readers more aware?

It is interesting to note that across the news or in a tangential way, the State and its people remain to be present in different capacities. Sometimes the state is directly involved, while at other times, indirectly so, but in every case, the State represents their people across the news. There are a number of cases where the State is absent and has nothing to do with the case but the onus lies upon the State to take cognizance of the issue and address the grievance. Therefore, overall, media plays an effective role in order to corner the State over various issues, whether implicitly or explicitly. In other words, the State remains in the locus of news or media. How is the State inevitable in the eyes of media? How does the State counter media?

[24]Ibid.: 71–2.

[25]Ibid.: 69.

1.2 Negotiating the State

The meaning of the modern State can be understood as being "an organization with de facto responsibilities for the government and administration of a territory. That territoriality is clearly a central feature of the geography of States, but geography is important in other ways too."[26] Merje Kuus and John Agnew argue, "the sovereign State, we posited, is not the basis for, but effect of, discourse of sovereignty, security, and identity."[27] Joe Painter infers Nicos Poulantzas's work that the state is an "institutional ensemble"—the mix of institutional forms and practices which Poulantzas calls the institutional materiality of the state. While further Painter infers Jessop, the State may be defined as "a specific institutional ensemble with multiple boundaries, no institutional fixity and no pre-given formal or substantive unity."[28] The State is an agglomeration of different institutions and each institution is supposed to work independently without overlapping and they work in a very cordial manner to support the State in its day-to-day activities.

Furthermore, the meaning of the State is never static, but changes with the changing dimensions of society. The meaning of the State has remained in use for political purposes. In the past, the State business was run by the "*pastor*" community or some equivalent religious authority. In the Islamic world, the administration was run by the "*Khalifa*"[29] or the Ottoman Empire,[30] which was later abolished by the British. With the increased mobility of people from one place to another, the people came to admire the information of best practice regarding the administration of their physical space (nation/country). As examined by Joe Painter, the idea of the modern State evolves from "the products of specific social processes and political struggles."[31] Joe Painter further argues that the theory of constant struggle between the powerful and powerless people over their dominion and further explains that "the process of State formation was a by-product of other activities, which may in themselves have been intentional, but which was not intended to generate the multifunctional modern State." What Anthony Giddens argues as "modern State formation was the 'unintended consequence' of intentional activities."[32]

The State continued to be conceived as being powerful and was supposed to assert power on their inhabitants through available mechanisms or institutions[33]. Of course, there are several questions which are still unanswered regarding the formation of the State (universally accepted meaning of the state, why the State been

[26]Painter (1995).

[27]Kuus and Agnew (2008).

[28]Painter (1995, p. 64).

[29]In Islam, Khalifa was custodian and ruler of Islamic territory, but abolished by Britishers.

[30]Istanbul (Turkey) is known for sitting place of Islamic rulers.

[31]Painter (1995, p. 35).

[32]Ibid.: 35.

[33]Dean (2010).

formed? What was the sole purpose? Did we achieve the purpose?), and shape (democratic, republic, socialist and welfare) of modern government and its function. How it was shaped to its present form and structure? Mitchell Dean also asks the question, "How did a secular rationality and art of government come about?"[34]

The monarch was the centre of power in the ancient period and was subsequently replaced to some extent by feudal lordship that existed alongside the monarch. The power enjoyed by the feudal lords was almost equivalent to that of the monarch in the concerned spatiality. The power struggle moved ahead with time and the nomenclature of power changed and incorporated modern practices became democratic, republic, socialist and welfare State over a period of time. These modes were adopted by particular nation according to their suitability. At the same time, there are still several countries across the world that continue with the rule of the monarch. However, the monarch mostly remains a symbolic head where the original power is vested in the parliament. For example, several European countries still follow monarchy such as Australia, Spain, the United Kingdom, etc.

With the establishment of the modern State, other institutions were formed to assist the functioning of the State. Amongst these institutions, some played a larger and catalytic role. In a democracy, the legislative, executive, judiciary and media are considered as the main pillars to sustain the State and to assist the State in moving ahead. These institutions have also clashed with each other over their power and jurisdiction. Judicial activism undermines the institutions for various purposes while media portrays the inefficiency of the institutions through various programmes.

In a democracy, each institution has its own position in the State. The institutions are supposed to deliver specific tasks in the interest of the State. For these institutions, the State is supreme. Except media, all three institutions (executive, judiciary and legislature) have no direct contact with the people of the country until and unless there are some issues or circumstances/provisions wherein they need to put forth their plea before the institutions. The institutions are supposed to take care for each individual's rights and lay down some criteria regarding their public/private matters when required. The only difference between media and these three institutions is that media approaches the people, while people are supposed to approach these institutions.

In contrast to the other three institutions, media is present everywhere in some form or the other. Media is "electrical" by nature. Its primary function is to retrieve information and disseminate that information through technology. Of course in between the retrieval of information, collection of information, sorting of information and sending/printing/ broadcasting/telecasting of information, a lot of processing takes place in order to reach the final destination. In the information age, media emerges as an important and nodal institution apart from the other institutions. Within the State, media is quite a new phenomenon and its expansion is the result of technical development and globalization. As Arvind Rajagopal explains:

[34]Ibid.: 101.

> Only twenty years ago, the total circulation of newspapers was 22 million and TV coverage reached 11 million households or about 55 million people. By 2006, TV coverage had expanded to 40 million households. By 2004, the media industry had become the fastest growing sector of Indian business, and nearly a fifth of the population lived in a television household. This represents an extraordinarily rapid increase. How can we understand the effects of this rapid growth? We might say, there are two models of globalization operating: one that conceives it as coming from above, and from without the country; and the other, as from within the country, albeit in interaction with forces from outside.[35]

The expansion of media is growing every day in India irrespective of its medium and uses. While the media business is growing, the consumption of media products is also growing at the same pace. Therefore, it is essential to know which medium of media is popular among people. How does it affect the citizen's views? How media do its business? What kind of content is produced by media? What is the relationship between media and the State? How media communicate with the State? How media influence politics?

As we know, media addresses its viewers through various mediums such as images, printings, audios, web pages, blogs, etc. The viewers might range from the State, institutions, organizations, groups, peoples, etc. The State, being the custodian of all institutions, exerts power through delegations when issues are supposed to be addressed in a figurative or pictorial form. Every issue can be identified with the help of a skeleton (overview idea of the issue through figures/pictorials) which provides a basic understanding in order to signify the background of the issue. The State and the media become intertwined by the issue, which is reported in various ways. For example, during the UP Assembly elections 2012, the State and the media confronted, represented and understood issues in different ways. But both shaped the issue during mutual interactions where issues were reshaped for political causes. Let us see how a State sometime adheres to media's attention, while others are unable to draw the same attention. The UP Assembly elections was quite remarkable, even though there were four other Indian states where elections were being held simultaneously; neither of them was able to draw the media's attention like the UP elections. The nature and dynamics of the UP elections were perhaps the reason behind the extensive media coverage given to it.

Why media had given UP election much importance in comparison with other states? How media being virtual space had discussed the real space? So, it is basic to understand that how media functioned as *space*.

---

[35] Rajagopal (2009, p. 17).

1.3 Theoretical Framework

Media and the State emerge as two distinct features as *space* when media posits as relational ways. While the State is space in abstract form where areal extent does matter and is supposed to shape media programmes at some extent. Berry et al. (2013) identified heterogeneous elements to mark media as space.

1.3.1 Media as Space

Berry et al. (2013) infers that "the presence of media in public space has transformed our understanding of both 'space' and 'publicness'."[36] He further added that "space as heterogeneous matter, comprised of diverse things and qualities, has become a philosophical concept worked upon in media and cultural geography, and in discourse of architecture and urban design. The significant value of thinking space through heterogeneity is twofold. First, heterogeneity posits difference as difference, not as deviation from a standard model (the town square, the badly lit alley) but instead as a proliferation of variegated spaces that overlaps and transmute. Second, heterogeneous space produced (rather than already existing) or, more explicitly, performed. The activities, events, small acts and transmissions in public space are a production of the environment, as Lefebvre argues forcefully in *The Production of Space*, which otherwise would not exist. Space, in other words, is imminent field of relations that are in constant flux as bodies, material forms and images come into contact."[37]

Chris Berry argues that "space emerges through practices that become consolidated in representational forms, such as tele-visual and filmic images or diagrammatic models. The unending need for space to be continually reproduced also needs to be understood here, a point that opens onto the possibilities of contestation; space is never finally fixed but only stabilized at certain historical moments. If it is the case, as has been argued, that enactments of space inevitably involve media, and then what we understand by 'media' in this scenario also demands consideration."[38]

Berry further said that "it is media that contributes to the 'atmosphere' of an environment in multiple ways. The affective capacity of media in public space is one of the least visible and yet most significant features of its presence. Media (as news streams, broken 'white noise' screens, recorded image loops montages, ambient sound and multi-screen live relays) create moods, tones and reminiscences, as discrete as the humidity and volume of the air and yet as charged as the weather. Interacting with particular conditions of a location, the meaning and experience of

---

[36]Berry et al. (2013, p. 5).

[37]Ibid.: 5.

[38]Ibid.: 6.

media is not knowable in advance; image, sounds and words are all changed by the contexts in which they appear."[39]

He identified heterogeneous images that are supposed to be used for different purposes, whether for public or private, to address the need of the spatiality therefore, he marked media as space where knowledge can be perceived and conceived in mundane ways which can stimulate the production of many spaces.

The production of space can be easily understand through the work of Lefebvre (1991) who argued that space and spatial relations are the material and social outcome of capital.[40] Appadurai (1996) puts forth the term "mediascapes" to understand the global culture wherein the role of media is inevitable and plays an important agent for transformation in various ways. He argues "scapes" to understand the flow of culture/information across the globe, the contemporary pattern supposed to influencing the worlds and helping to set an agenda as well.

Hence, the role of media in setting up an *agenda* is evident and it does so in order to shape and reshape the political discourse. Wu and Coleman (2009: 776) argue, "agenda setting theory is the phenomenon of the mass media selecting certain issues and portraying them frequently and prominently, which leads people to perceive those issues as more important than others. For more than thirty years, the main concept in agenda-setting theory has been the transfer of issue salience, or how media emphasis of certain issues raises their importance for the public."[41] However, apart from agenda setting, there are two more theories—"priming" and "framing." Scheufele and Tewksbury (2007) infer that priming occurs when news content suggests to news audiences that they ought to use specific issues as benchmarks for evaluating the performance of leaders and governments. It is often understood as an extension of agenda setting. Framing differs significantly from priming as it is based on the assumption that the way in which an issue is characterized in news reports can have an influence on how it is understood by audiences.[42]

Soja (1996) interpreted such image(s) or location as "Thirdspace." He further argues "thirdspace is a metaphor for the necessity to keep the consciousness of and the theorizing on spatiality radically open. It is used most broadly to highlight what are considered to be most interesting new ways of thinking about space and spatiality. In its broadest sense, it is purposefully tentative and flexible term that

---

[39]Ibid.: 6.

[40]Henri Lefebvre calls it 'representational spaces' in his book *The Production of Space* (1991: 39). On the foundations of first space, second space and the third space it can describe as a creative recombination and extension of spaces. The first space is about the "real" material world while second space is about the reality though "imagined" representation of spatiality and as proceeding into third space where "real" and "imagined" combined to represents the world. The real and imagined feature easily found across the media including audio-visual and print. However, Lefebvre do not directly acknowledge the media as virtual space rather he emphasis on the heterogeneity of space in myriad senses. Further, see Soja (1996).

[41]Wu and Coleman (2009).

[42]Scheufele and Tewksbury (2007).

attempts to capture what is actually a constant shifting and changing milieu of ideas, events, appearances, and meanings."[43] Herein, the space being discussed is lived space, a space that stretches across the images and symbols that accompany it. That is the "Thirdspace"—the radical enunciation of space. As Edward W. Soja argues, "the process of creating of space involves first and foremost direct and lived experience."[44] He further argues that "it is a space where universal issues like gender, sex, music, history, geography, sociology, politics, race, religion and class, etc. can be addressed easily from this platform without privileging one over the others; where one can belong to different ideology either Marxist, idealist, structuralist, and trans-disciplinary at the same time."[45]

Thirdspace is about to discuss the heterogeneous subjects irrespective of views, ideology and identity. Thus, we can assume that media is providing such platform wherein so many of images, identity and ideology are supposed to portrayed in myriad ways wherein role of media is inevitable.

Knut Lundby points out, "the Swedish media scholar Kent Asp in 1986 introduced a concept of 'mediatization' in a study of the power and influence of the media in political process."[46] He further argues that "the first social scientific works on mediatization were in the field of *politics*. How the media interfere and transform politics is continued and visible part of mediatization research. Mazzoleni summarizes the main effects of mediatization on politics: *first*, their agenda-setting capacity; *second*, that the media in general and television and the 'tabloid press' in particular make politics spectacular and personalized. The *third* effect is the fragmentation of politics; the *fourth*, the selection of political elites through the imposition of media-driven mechanisms on the political communication."[47]

1.4 Review of Literature

The literature review can understand through two ways: media and geography and image politics and its impact in everyday spatial activities.

1.4.1 Media and Geography

Jacquelin Burgess and John R. Gold in their work *Geography: The Media and Popular Culture* (1985) pointed out that geographers should give some attention

[43] Soja (2009).
[44] Soja (1996, p. 5).
[45] Ibid.: 5.
[46] Lundby (2009, p. 12).
[47] Ibid.: 6.

towards media in order to know literature, text, social and behavioural perspectives through media. They consider (media) as "an integral part of popular culture and, as such, are an essential element in molding individual and social experiences of the world that is shaping the relationship between people and place" (p. 1). The work comprises of ten articles and discusses media and its relevance in the concerned spatiality. Each article uses different approaches at the level of content and concept. While the first two articles are about international monitoring of the flow of programming and "how the 'unconscious categories' of highly complex television programming structure can be made explicit through the use of a microcomputer data base which classifies content in terms of different levels of generality" (p. 27). The other articles are about popular culture, examining the English landscape through documentary films, enquiring "hegemony and racism and nationalism in the Southern United States 1865–1900," "encoding newspaper texts" and "effects of media messages." The articles argue that "the rapid diffusion of these media has attracted the attention of a wide range of social scientists concerned about the impact that could result in terms of the organization and conduct of social, economic, and cultural life" (p. 6). However, they agreed that "over the last fifteen years, there have been various occasions on which geographers have acknowledged the importance of the media but, by and large, the quality of the ensuing analysis has been inadequate" (p. 5–6). The author tried to bring spatial glimpses across the world with the help of media that covered the discourse of spatiality into different dimensions.

Karin Ikas and Gerhard Wagner in their work *Communicating in the Third space* (2009) argue about the possible dimensions of third space from interdisciplinary approaches including "The location of culture," "The quest for inter- or transcultural communication," "The spatial turn," "Theorizing third space," "Literizing the third space" and "Locations and Negotiations." While beginning with perspectives that "at the turn of the 21st century, globalization has transformed the earth into a planet of nomads. Whether it is empires, states, nations, public spheres, religious communities, ethnic groups, townships, or neighborhoods, today million of migrants challenge the existing power structures and expose the available self-images as ideologies" (p. 1). The phrase "third space" was brought into the public domain by Homi K. Bhabha in his book *The Location of Culture*, wherein he exemplified that where two distinct cultures live side by side, they have the potential to negotiate and translate their cultural identity at some point, which might constitute the third space of enunciation. The author tries to put an understanding around third space in day-to-day socio-politico-cultural discourse wherein they are successful.

In a similar vein, Ulrich Beck and Britta Kalscheuer in the quest for inter- or transcultural communication highlight the ways in which cosmopolitan realism takes place across the world and how insecure they are at every ways. The thinking of the third space was further enhanced by Edward W. Soja and Julia Lossau through "The spatial turn" where both theorize about how spatiality has been discussed among the scholars in different capacities. Roberts J.C. Young took forward the thought in quite a comparable manner between Bhabha's notion of third space and Emile Benveniste's "Course in General Linguistic." Bill Ashcroft and Karin Ikas argue in literary discourse and look at Shakespeare's writing and characters to exemplify the idea of a third space. While Frank Schulze-Engler and Gerhard Wagner argue that looking at the third space beyond postcolonial constellations and internal conflicts makes a space in the territorial boundary in the country to raise enunciation debates. However, both the authors conceptualized the spatiality through multidisciplinary dimensions.

Knut Lundby in the book *Mediatization* (2009) argues that "Mediatization points to societal changes in contemporary high modern societies and the role of media and mediated communication in these transformations process of 'mediatization' affects almost all areas of social and cultural life in late modernity" (p. 1). The book was successful to raise the debate on theoretical grounds, "where the changes of mediatization come to the fore" (p. 14). The volume is divided into three sections. The first section examines the *concept* of mediatization, and the second section looks at major *changes* in society and culture where mediatization processes are involved while the third section was on consequences of mediatization in different arenas and aspects of social life (p. 14). Here, the author tried to establish the process of media effects that how media are supposed to work in day-to-day basis or to influence particularly in political contexts.

Maxwell E. Combs and Donald L. Shaw in their paper *The agenda-setting function of mass media* (2012) infer that "in choosing and displaying news, editors, newsroom staff, and broadcasters play an important part in shaping political reality. Readers learn not only about a given issue, but also much importance to attach to that issue from the amount of information in a news story and its positions. In reflecting what candidates are saying during a campaign, the mass media may well determine the important issues-that is, the media may set the 'agenda' of the campaign" (p. 153). They established the linkages that how media influence the election process, to mediate people, to follow media and to choose their political preferences as well.

Jeremy W. Crampton in his book *The Political Mapping of cyberspace* (2003) argues, "Cyberspace is a classic case of a space which is produced, and which in turn produces (spatialized) subjectivity. Thus not only can cyberspace be mapped, but in tracing out its contours, we are tracing out the lines on our faces. Cyberspace is an area of geographic knowledge that sits equally between society and technology. It invites the 'fieldwork' for critical politics of spatial representation and from that a critical politics of geography and space" (p. 6). The book has three parts: first part deals with spatial representation by analyzing how "one particular form of mapping, distributed interactive mapping, has risen to prominence historically in conjunction with digital communication technologies" (p. 7); second part deals with online identity crisis and how "cyberspace is undergoing a crisis of authentication" and; third part examines two case studies of "the production of the spatial politics in cyberspace" (p. 8). The author made explicit argument that the contents in the cyber space are purely spatial in nature and one can easily understand the nature of spatiality through available contents in cyber space.

1.4.2 Mapping Image Politics

Christopher Pinney in his research paper *The politics of popular Images: From Cow protection to M.K. Gandhi 1890–1950* Mahatama Gandhi, (2009) argues about the politics of popular images in India. The images either belong to a leader or something else that can generate politics. He notes that the politics of images usually circulates around the people in a political context. In India, the sacred is seen as a pure substance and creates a separate spatial identity when the "sacred symbol" and "sacred space" shape the image. To illustrate, he uses the image of a cow, leaders and paints these with a religious tinge to explain the politics of space. He further explains the image politics and dimensions of an image around cow. The cow is a sacred sign and symbol and of course, represents a society as well. The image of a cow forged a new sense of community amongst the people and the process of sanctification was carried to represent the sacred space in society. Pinney uses another example—during the twentieth century, Bengali Drama *Matri Puja* was seen as a seditious allegory and later various festivals were held to show its allegorical purpose. Pinney notes that the allegory may be different either in poem, plays and many other forms for critical purpose. While the author quotes Jean-Francois Lyotard about the "figure" figure, saying that "meaning is not produced and communicated, but intensities are felt." Lyotard also invoked the "figure" as being the opposite of discourse.

However, Pinney further examined the image of M.K. Gandhi and bhagat singh for various reasons. Both are well-known figures, but M.K. Gandhi was more accepted than Bhagat Singh across the global social space. The images of both are available in the market in vendor stalls or are seen hanging in offices and other public spaces and are quite distinguishable. The images of both leaders send two different messages across the social space. The image of Gandhi sends a message of sainthood or a polite man who believed in nonviolence, while the image of Bhagat Singh is in stark contrast to Gandhi's Brahmin identity depicted by him wearing the sacred thread. Bhagat Singh is seen twisting his moustache and carrying a pistol; a symbol of violence. Thus, the author places the image in two different directions—one is peace-loving and other believed in violent means.

Pinney further explores the image of Bhagat Singh with the help of a painting by Rup Kishor Kapur who was active in the Congress Party during the independence movement. The image of Bhagat Singh was supposed to depict martyrs at various points in history and later the image became symbolic of artefacts in the Indian culture. Rup Kishor Kapur painted a picture of Bhagat Singh when he was hanged. It shows Bhagat Singh beheaded, offering his head on a plate to Bharat Mata (India). Bharat Mata is shown weeping. Apart from these images, other celebrations of martyrdom were identified as being seditious. In another example, the writer also uses the image of Krishna along with the *sudarshan chakra*.

The depiction of Gandhi in Indian culture is almost that of an *avatar* (incarnation) of one of the Gods. Some of his images do reflect a divine Hindu identity. Pinney draws inspiration from the work of Ashis Nandy and argues that Gandhi was, in many respects as much Christian as Hindu and even in the "late 1940s the recurring theme is that of that parallelism between Gandhi and Christ and between Gandhi's assassination and the crucification of Christ". The depiction of Bose and Nehru as avatars also carried the image politics into new horizons and the image becomes a complete portrayal of the persona. Pinney interpreted the images of Gandhi and Bhagat Singh in two different contexts in order to understand the socio-politico spatiality. The relevance of both the leaders was inevitable during freedom movement in two different ways which Pinney had tried.

Arvind Rajagopal in his book *Politics after Television: Hindu Nationalism and the Reshaping of the Public in India* (2001), argues the ways in which television has profoundly changed the politics in India. The book examines the Ram Janamabhoomi Movement (RJM) and Babri Masjid, (BM) issue—which brought the BJP into the limelight and provided them with an opportunity to rule over the country and expand their political ideology as well. Rajagopal observes that Hindu nationalism shared their technologies and commodities (image) for expanding the market and reach the audience, respectively. He chooses Television as a medium to record the rhetorical episode which brought changes to both the people and politics in India. Television operates within the logic of capital exchange as well.

However media played a significant role in creating partisan identities. The division of people into two segments where one belongs to the English language while the others to vernacular languages, was one such division. The English language audience was relatively coherent, well networked, elite, while the Hindi language was larger in number and contrary to English language audience. The reach of Hindi media is more towards the interiors of the country than the relatively urban character of English language. The English press was largely limited to political and legal issues, while Hindi press was diverse in its coverage of social and political issues to reach out to their audience. The BJP utilized the Hindi press which was sympathetic to the RJM-BM.

Arvind Rajagopal examined the role of visual and print media that how media helped to shape RJM-BM issue for political goal. He argues that vernacular media was instrumental to radicalize the issue rather English media reported the issue in light of social tension that became wider.

S. Elizabeth Bird in her book *The Audience in Everyday Life* (2003) conducts an interdisciplinary study while trying to connect the relationship between Anthropology and media and echoes the way in which media has developed as culture in our social space. She is amazed about the audience and wishes to explore "how media articulate with such factors as class, gender, race, leisure and work habits, and countless other variables" (p. 4). She firmly believes that media is an extension of culture and the ethnography method will suit her (because she belongs to the discipline) to explore the culture in day-to-day affairs. In her chapters, she uses media discourse at length which surfaces in various forms but she examines each episode through the lens of an ethnographic approach. She endorsed media as part of culture that became inevitable in day-to-day activities like other cultural activities. How one can be influenced after media brief.

Wim van de Donk et al. in the book *Cyberprotest* (2004) acknowledges "social movements and ICTs' and argues for the use of ICTs during social movements. Meanwhile, it seeks to critically explore, analyze and assess the implications of the use of ICTs (such as internet) for "citizen mobilization and the formation activities, legitimacy and effectiveness of old, new and 'newest', social movements" (p. 1–2). The volume largely examines the field of ICTs and democracy. The study refers to the Internet and new media which has become ubiquitous in our day-to-day life and allows us to raise, seek and pursue our participation in social movements. The articles mainly deal with an approach about "how social movements are shaping the new media in ways that are both innovative and consistent with past traditions" (p. 24). The author examined the people's activism in social media and how social media is shaping the nature of protests under the surveillance of ICTs.

Pramod K. Nayar in the book *Digital Cool* (2012) examines "how everyday lives, identities, social relations, politics, leisure and realities are increasingly mediated by digital technologies, networked and programmable media" (p. 1). He sees media as cool technology and looks at how we use it in everyday life. As he says, "Digital Cool is the age of *simulation, hyper-realism, interactivity and hyper-mediated lives*" (p. 28). The book is interested in the social and cultural changes caused by "new media" where author point out that ICTs gadgets such as computers, the mobile phone and the Internet are playing an important role in an individual's life.

Paul C. Adams in his book *Geographies of Media and Communication* (2009) divides his book into four quadrants in which the first is media in space, second is spaces in media, third is places in media and the fourth is media in place. The first three chapters are about the history of media, the way in which evolution of communication takes place across the world. "Media in space" mainly deals with signals moving through the infrastructure in various paths and creates space as defined by communication traffic. "Spaces in media" can be understood as a "functional topological space," and "real in effects." "Places in media" constitutes not just words, pictures, and moving pictures, but also hybrids of all these. "Media in place" can be understood as when some communication becomes part of a place and that place would not be the same without them. The book follows an inter-disciplinary approach and examines the construction of the meaning of place and space within topological and communication pattern. The author makes a lucid attempt regarding the relevance of spatiality in both socially demarcated locations and more abstract constructs along with "places," "spaces," "network," and "topology." The book also discusses the spatial consequences of modern media and communication in order to define terms like "deterritorialization', 'cyberspace", and "network society", etc. Paul C. Adam tried to construct a spatiality through media contents that how media is supposed to create such virtual spaces and places in myriad ways.

Ather Farouqui in his edited volume *Muslims and Media Images* (2009) draws attention towards how media portrays Muslim images since the worldwide outrage against Muslims on media for harbouring terrorist's activities. The four part volume has the first part which deals with English media which includes different perspectives to show when and how things went wrong and who is responsible for the present stereotypical representation of Muslims on media. The second part looks at transcendence scholars who discuss boundaries where cultural outfits of Muslims appeared in print media. The papers exemplify both local and global perspectives. The third part deals with Urdu journalism and the working nature where Urdu newspapers are completely dependent on Western media or some other media and have not thought about their own source of news collection. All the papers in this section are sceptical about Urdu journalism and blame them for the negative portrayal of images of Muslims across the world. The fourth part is about popular images in films, the way in which Muslims are represented in Bollywood cinema. Author critically evaluated the media and their stereotypical representation of Indian Muslims in various shades.

Philip Hammond on his book *Media, War and Postmodernity* (2007) argues in the sense of "image, speculation and media presentation." The author looks at the Cold War period and the subsequent developments that took place in the Western world and examines the post 9/11 twin tower attacks. He finds the root cause for all this in literary language where language contains powerful meanings in modernity and postmodernity dualism. The first chapter of the book describes the idea that the "Western military and wars they fight can be described as 'postmodern'" (p. 11). The second chapter examines the "humanitarian intervention" where Western media is strategically used to settle the military score. The third chapter is on the war on terror where the role of media is critically examined. The fourth chapter deals with the post-Vietnam war conditions and cultural understanding that the war was fought because it involved different cultures. The fifth chapter echoes the way in which war creates security problems and "risk society" at different orders. The last chapter critically examines the narratives and evaluates the postmodern value where the "death of subject" is imminent. The author sees the war episodes of the Western world as cultural gaps and considered the role of media that is creating an atmosphere of war in various capacities whether humanitarianism, ideological, ethical or moral basis.

Joshua Meyrowitz in the book *No Sense of Place* (1985) discusses media and social change and how media as a mechanism cajoles social change and behaviour. The book exemplifies personal experience in order to understand the dimensions of place through in media. The first part of the book lays out a general framework for studying media and social change and incorporates approaches like "medium theory" (of McLuhan and others) "situationism" (of Goffman and others), "information systems", "behaviour" with reference to media and social behaviour. The second part discusses the "ways in which electronic media affect the information systems" and how electronic (picture) is more effective than word. The third part looks at the new social landscape where physical and social boundaries blur. The fourth part discusses the dimensions of social change on masculinity and femininity, childhood and adulthood and political leaders and their followers. The fifth part suggests that "placeless" as live culture. Author examined media and social change as demand and supply contexts wherein socio-behavioural changes are reflection of media consumption.

D.N. Rodowick in the book *Reading the Figural, OR, Philosophy after the new Media* (2001) examines the "figural" and its discourse, predominant over cinema, photograph and media. In the first chapter, the author discusses Lyotard's "concept of *Discourse, figure* as well as his aesthetic." The second chapter involves a philosophical discussion to maintain the "figural" as experiential and mobile concept, although using Deleuze and Michel Foucault's reading on power. The third chapter deals with contemporary film theory where two approaches "Thierry Kunzel's concept of figuration, or semiotic "constellation" in the film work, and Marie-Claire Ropars-Wuilleumier's theory of *cinecriture*," are discussed (p. xii). The fourth chapter is about the philosophical reading of "Kant's Critique of Judgment as played out in Jacques Derrida's book Truth in Painting and his essay 'Economimesis'" (p. xiii). The fifth chapter is about the "figural as a historical concept in the idea of 'historical images' elaborated in Walter Benjamin's and Siegfried Kracauer's studies of film and photography" (p. xiv). The sixth chapter deals with examining Deleuze's theory of film in the context of philosophy of history. Rodowick develops the discourse on the *figural* based on philosophical and historical contexts where the subsuming of power was interpreted in complex language. The book reflects an interdisciplinary approach wherein different theories were put all together to understand that how media works.

Gaurav Desai's edited volume *The Virtual Transformation of the Public Sphere* (2013), is an outcome of a seminar proceeding where different scholars put words on the dynamics of the virtual and public sphere concept of Habermas and how it became part and parcel of an individual's life from providing information or keeping everyone in touch through a click. The articles in the volume involve using a theoretical approach, Internet connectivity, the virtual politics, virtual sexuality and virtual pedagogy. Each article carries the Habermas notion of "public sphere" and substantiates the growing of virtualise views in myriad forms. The author discusses how virtual(ism) takes place in an individual's day-to-day activities and even cajoles them at various degrees.

Oliver Boyd-Barrett and Peter Braham in the edited volume *Media, Knowledge and Power* (1987) argue that "the role of the communication media in the production and social distribution of knowledge and in the process whereby individuals learns from media" (from last Para of editor's introduction). The author argues, "the three dimensions of communication: the nature and use of language, the 'mediation' of communication by technologies such as print, television and computers, and the social organization of communication among professional groups" (ibid.). The companion volume deals with "media studies" in the first two sections, while in the next two sections, the authors move towards the education imparted by communication media. The next two sections are about information studies and the ongoing research in the field. The author tried to balance media content and education in order to support the arguments of communication media as a centre of knowledge production and one that is explicitly bounded by power intricacies.

Masaaki Kimura and Akio Tanabe in the volume *The State in India: Past and Present* (2006) looking at the past and present contexts to trace the genesis of the state in India and how the shape of the state has been transformed since its inception in India. The first section of the volume is about the formation and concept of the state, the second section discusses "forms and process of the state" and the third section is about "ideas and problematics of the state." The book objectively discusses the past and present structure of the state in India. The author acknowledges the existence of various ideologies that surface over the ash of the Cold War as "religious nationalism, ethno-nationalism and fundamentalism" that bore ethnic, religious, gender, linguistic, regional, sexuality and caste identity in order to make a separate entity in the coexistent body. The author considers these identities both as "crisis" and as "representation," which moved in an entwined manner influencing the state at best while the author considers these act as problematic for the state and proposes that the state should come out heavily on these "problematic" of the state.

David Finkelstein and Douglas M. Peers in their work *Negotiating India in the Nineteenth-Century Media* (2000) discuss the functioning of media and media history particularly print media history during the Colonial period, and the way in which print media starts its sojourn in India as vernacular press. The book sheds light on media and its role during the period and contemporary thought regarding the same to examine the history of India. The writers wish to examine the general perception of Victorians who are usually seen as exploiters. The author argued "the result is an engagement of discourses reflecting not a consensual view of India, but a fractured one: a multiplicity of voices within and outside Indian borders, real and imagined borders, many interpretive communities from which individuals shift and slide to, from and in between, contesting cultural, religious, ethnic and gendered identities, shaping and being shaped through rhetoric and debate" (p. 19). The book examines the vernacular and English newspapers during the nineteenth-century and compares the way in which growth its takes place parallel to the Western world particularly England under whom India was being shaped by media.

Per Stahlberg in his book *Writing Society through Media* (2013) performs an ethnographic study of a Hindi daily newspaper *Dainik Jagran*, which is published from Lucknow. He exclusively covered the day-to-day news about the city and the surrounding areas. "What kind of society is emerging from the impression one gets by browsing through a Lucknow newspaper?" (ibid.: 114). He believes that newspaper is a medium to map society in various ways.

Manish Tiwari and Rajan Pandey in his book *Battleground UP: Politics in the Land of Ram* (2013) exclusively cover the UP Assembly elections 2012 and comment on the politics during the election campaign in the state. The book comprises of six parts and each part discusses the following broad themes: the rally; war room and open fields; why Jambo (Elephant) crashed; politics of social engineering; image of average UP politician and; image of the Gandhi family and others. The authors tried to understand the region through political differences. The book provides political glimpses of political campaign during UP Assembly election 2012 that how and what necessary political actions are taken in order to make political score on opponents.

The relation between politics and media is inevitable and across disciplines one can observe that both politics and media are dependent to each other. For media and politics, political activities are like a commodity and supposed to disseminate the political information. The above mentioned literature review offers an interdisciplinary perspective of media and the state as well as how media and the state interprets the issues with spatial contexts.

1.5 Focus of Study

This study is about to know the answer of some of research questions that: To understand that how media space influences the state and; how the state and media space shapes territorial imagination and the underlining politics of image construction as well. Moreover, there are some research questions: To examine the nature of media in India; To examine how media constructed UP as the "heartland" of Indian Politics and Polity; To analyze political news and how it is shaping and reshaping the social space of UP, and at last; To understand how political news constructed the image of UP.

1.6 Study Workout Plan

This study involves both qualitative and quantitative techniques and considered news regarding UP Assembly election 2012 as case study which appeared across the media during the election in various forms. The study used both primary and secondary sources of data. The primary sources of material are the news contents in shape of e-news, print news and audiovisual (TV) news. News[48] has been collected from two different sources of media-one from newspapers and the second from TV news channels in both Hindi and English language (Table 1.1).

The study considers three newspapers—two from English (*The Times of India* and *The Hindu*) newspaper and one from Hindi (*Rastrya Sahara*) newspaper. *The Times of India* was selected on the basis of circulation while *The Hindu* was selected in order to know about the left driven ideologue on news or an image. However, the Hindi newspaper *Rastriya Sahara* was selected on the basis of its local affiliations (it is a UP based newspaper which is published at four locations in the state) and is popular in the region. *Rastriya Sahara* also runs an Urdu daily newspaper in the state.

[48]Stephens (2007) defines news "as new information about a subject of some public interest that is shared with some portion of the public" (quote in Nagpal 2011: 23). Journalists tell people the things they don't know about and help gain a balanced perspective about things they already know. Besides providing information and aiding learning, news also increases awareness of our societal norms, value systems and beliefs. News brings us closer to people of other cultures (ibid.: 23).

Table 1.1 Selected newspapers and TV news channels

	Newspaper	TV news channels
Hindi	1. Rastriya Sahara	1. Aaj Tak 2. DD 3. Zee news
English	1. The Times of India 2. The Hindu	1. NDTV 24*7 2. CNN IBN

The study has also looked at five TV news channels—three from Hindi news channels and two from English news channels. The Hindi news channels are; Aaj Tak, Zee news and DD (Doordarshan) while the English news channels are; NDTV 24*7 and CNN IBN. These news channels have been considered on the basis of their popularity. However, there is no uniformity in TRPs, as each channel makes its own claims regarding such ratings. Hence, these private news channels were given preference over other news channels because they are considered as the early TV news channels that revolutionized the TV news industry in India. The study has also considered DD news channel to acknowledge the state's views. This study has taken both Hindi and English language news content into consideration for two reasons - one is that Hindi is our National language and the major spoken language in the country and the second is that the English language is considered to decode the thinking of the literate people. However, the study collected news content from October 1, 2011 to until the declaration of the UP Assembly election 2012 result (March 6, 2012).

The secondary source of data was collected from; centre for the study of Developing Societies (CSDS), Internet and Mobile Association for India (IAMAI), Television Audience Measure (TAM)-India and from Ministry of Information and Broadcasting, Government of India. The 16th UP Assembly election 2012 data were collected from CSDS. The data related to people's engagement in social media were collected from IAMAI while about television user's related data were collected from TAM-India and the current status of both public and private media were collected from Ministry of Information and Broadcasting.

The study hold three focus group discussion (FGD)-each focus group comprises 5–7 peoples who belongs to UP-first those people who born and brought up in UP and second who born and brought up in UP but politically active and living outside the state for livelihood in Delhi. The focus group members were asked to make open comment on the issues raised by media (Appendix B). The focus group peoples were selected from different sections of the society including Youth (above 18 years), Gender (both male and female), Political activists (representation of all political party were included), Professional (working professional jobs in both public and private institutions were included) and Religion (peoples from all religion representation were included).

The news content was collected in shape of headline,[49] caption[50] and photograph[51] that was carried by the newspapers and TV news channels to report the political discourse around an image. The news headline is a medium which describes an image rather than waiting to summarize the complete story. The dimension of a caption is an inevitable part of news and the appropriate use of captions shapes the contexts of news and helps an individual to interpret the news in various contexts. The photograph/visuals are an appealing tool to analyze any image in any direction because people generally forget the headline and caption but after looking at the photographs/visuals they are reminded of the incident. FGD held and each member of the group was interview with structured questionnaire.

This study used content analysis[52] as a tool to understand the contexts of texts and visuals appeared in the different sources including media as well. The study collected news items in eight broad election themes (Appendix A) which appeared across the media during the election. The news items were collected and classified as media coverage of news (Table 3.7), number of news appeared in Hindi and English language (Table 3.8) and number of news appeared in selected media spaces (Table 3.9).

The content analysis used to interpret the texts and photos appeared in news even without distorting the actual interpretation at best. Therefore, semiotic

---

[49]"The functions of headlines are to attract viewer attention, depict the mood of story and index the contents of news bulletin. Hence the headlines must be crispy and attention grabbing. But the serious nature of news should be maintained except in soft stories. Headlines also encourage viewers to watch an interesting segment in anticipation of interesting news which is on the way" (Nagpal 2011: 24).

[50]"The caption is easy: just describe the photo or image succinctly and accurately. But when captioning photos for newspapers, captioning becomes more of an art form. Captions—or *cutlines*, as they're frequently called—must, like a good headline, draw readers into the story. Cutlines are full sentences, sometimes two", (http://facstaff.uww.edu/carlberj/230/hpc.htm. Accessed, April 23, 2013).

[51]Photographs generally work as memory weapon when people are supposed to grasp the news in the concerned contexts. Therefore, photographs are considered important elements in news compiling.

[52]Content analysis is about analyzing media content in systematic ways wherein "sample of media, establishing categories of content, measuring the presence of each category within the sample, and interpreting the results, usually against some external criteria" (Bertrand and Hughes 2005, p. 198). This study will follow the methods of content analysis to interpret the headlines, caption and photographs. Content analysis (Mitchell 2011; Bertrand and Hughes 2005) is a research tool focused on the actual content of media. To conduct a content analysis on a text, the text is coded or broken down, into manageable categories on a variety of levels—word, word sense, phrase, sentence, or theme—and then examined using one of content analysis' basic methods: conceptual analysis or relational analysis. It is used to determine the presence of certain words, concepts, themes, phrases, characters, or sentences within texts or sets of texts and to quantify this presence in an objective manner. Texts can be defined broadly as books, book chapters, essays, interviews, discussions, *newspaper headlines* and articles, historical documents, speeches, conversations, advertising, theatre, informal conversation, or any occurrence of communicative language. Krippendorff (2004). See for further details *Content Analysis: An Introduction to its Methodology* (2nd ed.). Thousand Oaks, CA: Sage. Neuendorf (2002), Weber (1990), Krippendorff and Bock (2008).

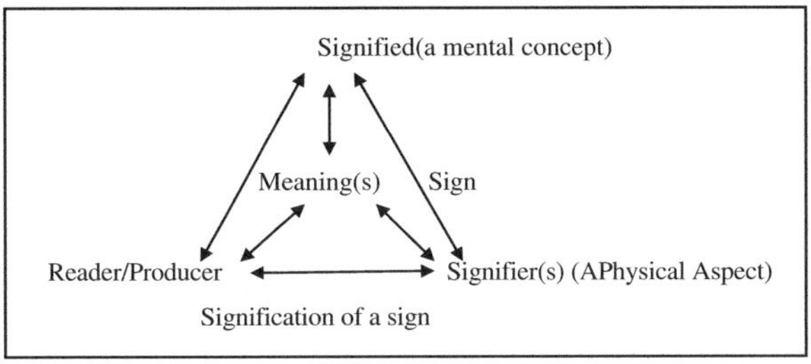

Fig. 1.1 Semiotic analysis

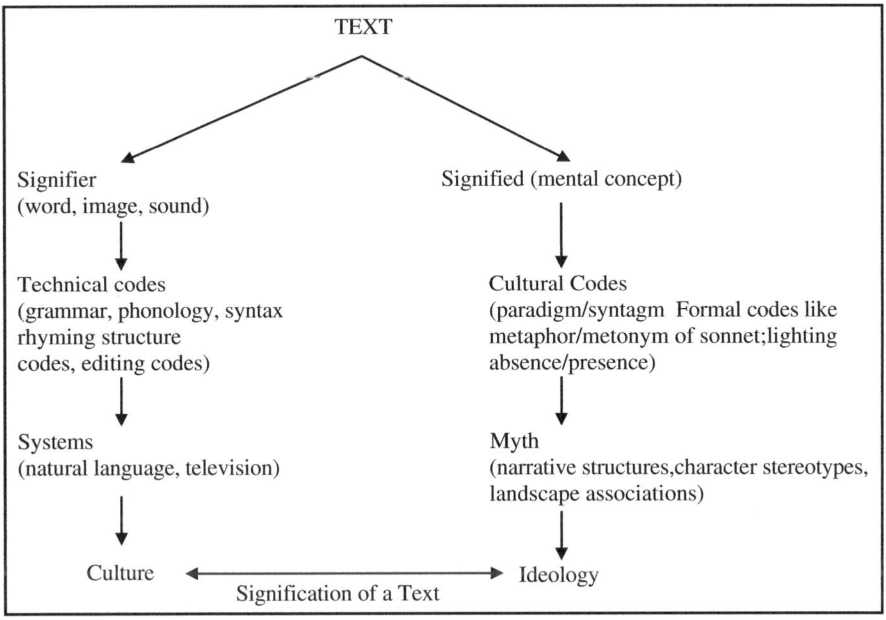

Fig. 1.2 Text interpretation

analysis[53] (Fig. 1.1) is used to understand the actual meaning of photos appearing across the media.

News is a kind of text which need clarity of expression in order to reflect the actual content in a lucid manner therefore it needs a systematic process to follow news. However, texts appeared in the media can be understood or can interpret through the following model (Fig. 1.2).[54]

[53]Semiotic analysis is about the study of signs and symbols. For more details see, Bertrand and Hughes (2005, p. 203).

[54]Bertrand and Hughes (2005, p. 204).

How media contents supposed to influenced an individual or in other word, media effects on individual (Fig. 1.3).

However, the basic problem of this study was that news items were often contentious in nature and some extent eroding the nature of texts. This study comprises of news issues which were generally found in inflammatory and considered contentious in public life. The considered news items were often derogatory. Therefore, the news items were further restructured and toned down in order to reflect the actual meanings without distorting the actual contexts. Particularly, the Hindi news items were carefully tone down in order to show the actual intentions in some places at best.

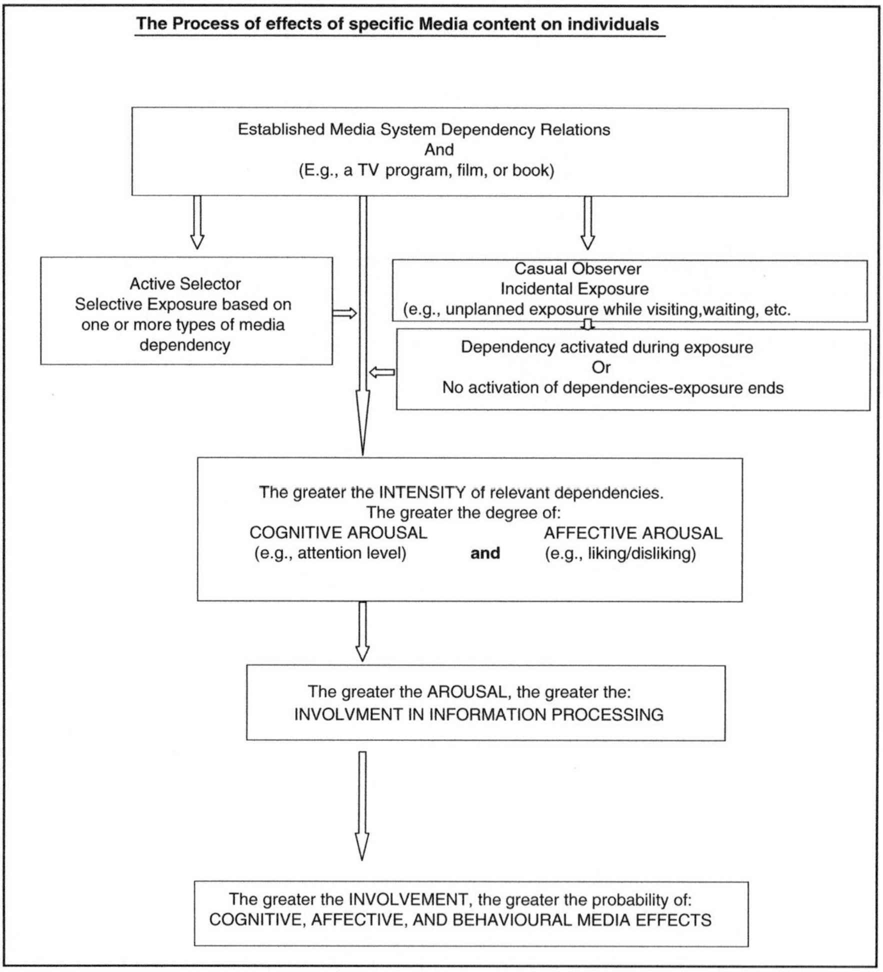

Fig. 1.3 Media effect. *Source* Defluer and Ball-Rokeach (1989)

1.7 Relevance

We live in a democratic set up wherein elections supposed to hold in regular/interval features to celebrate as political festival. In a democracy, media and state work together. During elections, the role of the media is that people want to know the news of "other" regions/spaces as well. This study aims to reveal the ways in which media supposed to represents the political news in their respective media spaces. What political identity/ideology media supposed to attached with? How does media maintain its relationship with state? How does a particular image become an identity? How one section of the media is different from another section when both are supposed to be reporting the same image? How media does shaped and reshaped the image politics and how and at what extent media and the state interpreted the image politics on political image? It is generally considered that who control UP politics shall control centre politics and considered "Heartland" of Indian politics as well. This study would have move through eight issues (images), collected from media (Zee News). How did an image become the centre of politics and how did politics play around an image during the elections? How was an image constructed as a symbol to represent the space/region?

1.8 Conclusion

This study is to understand the complex functional mechanism of media and the state during election time and at what extent both use image(s) as product. This study used UP Assembly election 2012, as case study and examined the role of media and the state that at what extent they can influence or determine the outcome of the election. Theoretical framework is conceptualized to understand the dimensions of both media and the state; the work of Chris Berry, Arjun Appadurai, Marshall McLuhan, Denis Wu and Coleman, Edward W. Soja, H. Lefebvre, Knut Lundby and Kent Asp are collected to sketch the study. The literature review posits stimulation and provided another direction to map the study at best. The study based on two research questions and four research objectives as well as both primary and secondary sources are used through content and semiotic analysis.

The next chapter is about the history, growth and development of Indian media. The contribution of old and new media is witnessed in mundane shapes whether during Colonial India or after independence or the expansion of social media in the country. The reach of social media is observed across different sections of people and at what ways old and new media shaped the political economy in the country.

References

Adams PC (2009) Geographies of media and communication: a critical introduction. Wiley-Blackwell, Chichester
Appadurai A (1996) Modernity at large: cultural dimensions of globalisation. University of Minnesota Press, Minneapolis
Barrett Oliver B, Braham P (1987) Media, knowledge and power. CROOM HELM, London
Berry C et al (2013) Public space, media space. Routledge, New York
Bertrand I, Hughes P (2005) Media research methods: audiences, institutions texts. Palgrave Macmillan, New York
Bird SE (2003) Audience in everyday life: living in a media world. Routledge, New York
Burgess J, Gold RJ (1985) Geography the media and popular culture. Croom Helm, London
Couldry Nick (2012) Media, society, world: social theory and digital media practice. Polity, Cambridge
Crampton JW (2003) The political mapping of cyberspace. Edinburgh University Press, Edinburgh
de Certeau M (2009) The practice of everyday life. In Thornham S et al (eds) Media studies: a reader, 3rd edn. Edinburgh University Press, Edinburgh
Defluer LM, Ball-Rokeach S (1989) Theories of mass communication, 5th edn. Longman, New York
Dean M (2010) Governmentality: power and rule in modern society. Sage Publications, Los Angeles, p 101
Desai G (2013) The virtual transformation of the public sphere. Routledge, New Delhi
Durham MG, Kellener DM (eds) (2001, reprint 2005) Media and cultural studies: key works. Blackwell Publishing, Oxford
Farouqui A (ed) (2009) Muslims and media images: news versus views. Oxford University Press, New Delhi
Finkelstein D, Peers M (2000) Negotiating India in the Nineteenth-Century Media. Macmillan Press Ltd, New York
Gorman Lyn, Mclean David (2009) Media and Society into the 21st century. Wiley-Blackwell, West Sussex, p 230
Habermas J (1989[1962]) The structural transformation of the public sphere: an inquiry into a category of bourgeois society. Polity Press, Cambridge
Habermas J (2010) The public sphere: an encyclopedia article. In: Greer Chris (ed) Crime and media: a reader. Routledge, London
Ikas K, Wagner G (eds) (2009) Communicating in the third space. Blackwell, New York
Kittler F (1999) Gramophone, film, typewriter. Stanford University Press, Stanford
Krippendorff K (2004) Content analysis: an introduction to its methodology, 2nd edn. Sage, Thousand Oaks
Krippendorff Klaus, Bock MA (eds) (2008) The content analysis reader. Sage, Thousand Oaks
Kuus M, Agnew J (2008) Theorizing the state geographically: sovereignty, subjectivity, territoriality. In: Cox KR et al (eds) Political geography. Sage Publications, Los Angeles
Lefebvre H (1991) The production of spaces (Donald Nicholson-Smith Trans.). Blackwell, Oxford
Lundby K (2009) Mediatization: concept, change consequences. Peter Lang, New York
McLuhan M (1964) Understanding media: the extensions of man. Mc Graw Hill, New York
McLuhan M (2006, reprint) The medium is the message. In: Durham MG, Kellner DM (eds) Media and cultural studies: key works. Blackwell, London
Meyrowitz J (1985) No sense of place: the impact of electronic media on social behaviour. Oxford University Press, New York
Mitchell WJT (1992) The pictorial form. Artforum 30:7
Mitchell C (2011) Doing visual research. Sage Publications, Los Angeles
Nagpal A (2011) From news to infotainment to tamasha: from tamasha back to news. The Publisher, New Delhi
Nayar PK (2012) Digital cool: life in the age of new media. Orient Blackswan, New Delhi

Neuendorf Kimberly A (2002) The content analysis guidebook. Sage, Thousand Oaks, CA
Painter Joe (1995) Politics, geography, and political geography: a critical perspective. Arnold, New York and London, p 32
Pinney C (2009) The politics of popular Images: from cow protection to M.K. Gandhi 1890–1950. In Rajagopal A (ed) The Indian public sphere: reading in media history. Oxford University Press, New Delhi
Rajagopal A (2001) Politics after television: Hindu nationalism and the reshaping of the public in India. Cambridge University Press, Cambridge
Rajagopal A (2009) Indian public sphere: readings in media history. Oxford University Press, New Delhi
Ranciere J (2007) The future of image. Verso, London
Rodowick DN (2001) Reading the figural, or, philosophy after the new media. Duke University Press, Durham
Rose G (2003) On the need to Ask how, exactly, is geography "Visual"? Antipode 35(2):212–221
Scheufele DA, Tewksbury D (2007) Framing, agenda setting, and priming: the evolution of three media effects models. J Commun 57:9–20
Shrivastava KM (1998) Media towards 21st century. Sterling Publishers Private Limited, New Delhi, p 1
Soja E (1996) Third space: journey to loss Angles and other real and imagined places. Blackwell, London
Soja EW (2009) Thirdspace: toward a new consciousness of space and spatiality. In: Ikas Karin, Wagner Gerhard (eds) Communicating in the third space. Blackwell, New York, p 50
Stahlberg P (2013) Writing society through media: ethnography of a Hindi daily. Rawat Publications, Jaipur
Stephens M (2007) A history of news. Oxford University Press, New York
Thornes JE (2004) The Visual Turn and Geography. Antipode 36:787–794
Tiwari M, Pandey R (2013) BattleGround UP: politics in the land of Ram. Tranquebar, Chennai
Weber RP (1990) Basic content analysis, 2nd edn. Sage, Newbury Park
Wu HD, Coleman R (2009) Advancing agenda-setting theory: the comparative strength and new contingent conditions of the two levels of agenda setting effects. J Mass Commun Q 86 (4):775–789 (winter)

Chapter 2
Mapping the Mediascapes in India

Abstract Media being the representational space(s) produces and disseminates information and knowledge in mundane shape and size; interpretation and representation. Media played as catalytic role during Colonial rule to unite people against Britishers or after independence to improve agricultural production, sex ratio, literacy rate, etc. through all medium-print, radio and TV and now social media in the country. Both old and new media witnessed: in different proportions at both rural and urban sphere, in all age groups irrespective of gender and, as platform for social movement in the country.

Keywords Mediascapes · Social media · Space · Place · Political economy

Arjun Appadurai infers that the concept of "scapes" "recognizes communication flows as structure and as ways of knowing the world, permitting deeper analysis of virtualization and globalization process." He uses five "perspectival constructs"

S. Moinuddin, *Mediascape and The State*, Springer Geography,
DOI 10.1007/978-3-319-51932-6_2

including "mediascapes"[1] ethnoscapes, technoscapes, financescapes, and ideoscapes.[2] Each "scapes" are basic steps towards understanding how the flow of communication takes place and further points out that mediascapes forms and informs through its "flows of images and narratives about human lives, as well as the technologies disseminating such images, including newspapers, magazines, television, and film."[3] Every day media acknowledges spaces and sometimes gives them a new identity. In the twenty-first century, media is considered an inevitable asset in both the developed and developing world. However, the use pattern of media varies across the spaces and an individual consumes media daily in various ways, just like eating every day, reading or listening or watching media are considered parallel acts although they differ in nature and characteristics. Media consumption is growing everyday wherein one kind of media is identifiable from another and shaping business interests in the country.

2.1 Mapping the Landscapes of Media

What is mapping? Mapping is about politics while the "politics is all about assumptions; perceptions of interest and values. There is therefore a tension in its mapping."[4] Mapping is an act of representation. It is about creating a political identity and what critical cartographers Roger Downs and Lynn Liben argue,

[1]"Mediascapes refer both to the distribution of the electronic capabilities to produce and disseminate information (newspapers, magazine, television stations, and film production studios), which are now available to a growing number of private interests throughout the world, and to the images of the world created by these media. These images involve many complicated inflections, depending on their mode (documentary or entertainment), their hardware (electronic or pre-electronic), their audiences (local, national, or transnational), and the interests of those who own and control them. What is most important about these mediascapes is that they provide (especially in their television, film, and cassette forms) large and complex repertoires of images, narratives, and ethnoscapes to viewers throughout politics are profoundly mixed. What this means is that many audiences around the world experiences the media themselves as a complicated and interconnected repertoire of print, celluloid, electronic screens, and bill-boards. The lines between the realistic and the fictional landscapes they see blurred, so that the farther away audiences are from the direct experiences of metropolitan life, the more likely they are to construct imagined worlds that are chimerical, aesthetic, even fantastic objects, particularly if assessed by the criteria of some other perspective, some other imagined world. Mediascapes, whether produced by private or state interests, tend to be image-centred, narrative-based accounts of strips of reality, and what they offer to those who experience and transform them is a series of elements (such as characters, plots, and textual forms) out of which scripts can be formed of imagined lives, their own as well as those of others living in other places. These scripts can and do get disaggregated into complex sets of metaphors by which people live as they help to constitute narratives of the Other and protonarratives of possible lives, fantasies that could become prolegomena to the desire for acquisition and movement" (Appadurai 1996: 35–35).

[2]Adams (2009, p. 112).

[3]Ibid.: 112–13.

[4]Black (1997, p. 105).

"creates specific spatial knowledge and meanings by identifying, naming, categorizing, excluding, and ordering."[5] Mapping is about the demarcation of power and contestation of the power. Mapping is a political tool wherein each political discourse analyzes its knowledge and meaning in particular ways.

Media is inevitable to society. As Knut Lundby points out, "Everyday life and the operation of society depend on mediated communication."[6] Modern media, which is a combination of old and new media, was developed as a sign of technological growth in order to show spatial glimpses which manifested themselves in the form of *images*. These manifestations came through a mixed blend of "real" and "imagined" portrayals or representations. Since its inception, media acquired social recognition and became a social phenomenon in terms of influencing and determining the spatial course of action. Media is an omnipresent phenomenon. Each country, whether developed or developing, has some sort of media practices and uses media gadgets to be in touch across the world. Both the rich and poor use media in different ways. The pulse of our everyday life is constantly interrupted by media in various ways. It interferes and impinges upon both our public and private lives. How much does an individual use media in their daily lives? Jeffrey and Doron (2012) offer new insights on how the BSP (Bahujan Samaj Party) used mobile phone technology to gain power and win the Assembly elections in 2007 in India. Thus the extent to which media can influence our lives cannot be undermined.

Media is a comprehensive idea that includes electronic gadgets ranging from old gadgets—telegraph, print, radio, cinema, etc. to new ones—TV, smart mobile phone, iPod, Internet, etc. "New media technologies are transforming the place of the media in home and life and are contributing to the changing 'where' of the everyday. Increasingly, the pulse of the everyday, beating in streets, in workplaces, in transport hubs, in cars, trains, planes and buses as well as in homes, is multi-layered and on the move, happening simultaneously here *and* there, or operating across thick global transport and communication connections."[7]

Habermas developed the idea of the "Public Sphere" and focused on the inevitable role of media in democratic administration. McLuhan (1964) further inferred, "The Medium is the Message" and explored the role of media in order to explore the ways in which media shapes an individual's consciousness, culture and society, and is considered as an active agent of historical change. Chomsky and Herman concluded their Marxist analysis of media as "Manufacturing Consent" through the influence of powerful individuals who are the heads of these institutions and who work collectively to maintain the interests of influential people in a class society. Hall's "Encoding-Decoding" offers to understand not just elements of media communication, but the communication process as a whole, elucidating the ways in which communication is retrieved and sent. Baudrillard's "Simulacra and

[5]Crampton (2010, p. 45).

[6]Lundby (2009, p. 1).

[7]Thornaham et al. (2009).

Simulation" is a postmodern vision of a world in which media represents the everyday events to such a minute extent that the boundaries between the "represented" and the "real" are blurred, producing a mediatised "hyper-reality" where simulations, signs and codes come to constitute everyday activities.

Henri Lefebvre in his book *The Production of Space* (1991) argues about representational spaces and examines the "embodying complex symbolisms, sometimes coded, sometime not, linked to the clandestine or underground side of social life, as also to art."[8] He further argued

> Space as directly *lived* through its associated images and symbols, and hence the space of 'inhabitants' and 'users', but also some of artists and perhaps of those, such as a few writers and philosophers, who describe and aspire to do no more than describe. This is dominated-and hence passively experienced-space which the imagination seeks to change and appropriate. It overlays physical space and making symbolic use of its objects. Thus representational spaces may be said, through again with certain expectations, to tend towards more or less coherent systems of non-verbal symbols and signs.[9]

Representational spaces are experienced in our daily lives through symbols of the "real" and the "imagined" and posed as being alive. Lefebvre explains

> Representational space is alive: it speaks, it has an affective kernel or centre: Ego, bed, bedroom, dwelling, house; or: square, church, graveyard. It embraces the loci of passion, of action and of lived situations, and thus immediately implies time. Consequently it may be qualified in various ways: it may be directional, situational or relational, because it is essentially qualitative, fluid and dynamic.[10]

The issues raised and discussed extensively during the election period in Uttar Pradesh (India) had to some extent shoved spatiality into something directional, situational and relational with spatial expectations. Of course, these issues are set to exemplify the spatial manifestations. For example, during the Assembly elections, when issues were discussed across media, they became situational, relational and directional, wherein every issue had its own significance either for a short period of time or for a long duration. The issues were discussed across media in such a way that every issue surfaced as representational spaces. As Lefebvre points out, "Representational spaces are symbolic works. These are often unique; sometimes they set in train 'aesthetic' trends and, after a time, having provoked a series of manifestations and incursions into the imaginary, run out of stream."[11] He further notes, "Representational spaces, which are limited to works, images and memories whose content, whether sensory, sensual or sexual, is so far displaced that it barely achieves symbolic force."[12] Hence, the election issues are merely symbolic in nature in order to raise expectations. The content of issues to some extent is

[8]Lefebvre (1991).

[9]Ibid.: 39.

[10]Ibid.: 42.

[11]Ibid.: 42.

[12]Ibid.: 50.

achieved and delivered through verbal, symbolic and sign: a medium of manifestations. Let us consider that media is seen as a representational space, because of its programme content, politics, ideology, version, reflection, and being *lived* in nature, and through symbols and images, making spatial inferences. Media adheres to the new and old mediums in order to serve its purpose in complex, straight and binary formations where issue is shaped, to a large extent, through comparison.

Representational spaces include the media space produced in order to subscribe to an ideology, relation, direction and consumption with respect to different contexts and in a "real" and "imagined" order. Media becomes a symbol of space and time in the *lived* sense in a cultural context where things and ideas are a mix of the "real" and "imagined." In what ways media produces an image after the mixing up of the real and imagined elements? What is the purpose of this intermixing? How do the produced images become imbued with political meaning? How constructed image becomes a symbol? In what way media helps in furthering politics? There are some questions that the study seeks to understand.

Media being a product/medium encourages users to make the usage of media including social networking sites (Google+, Facebook, Twitter, ibibo, Hangout, Hike and Whatsapp.), print (newspaper, magazine, etc.), audio (radio, FM, etc.) and audiovisual mediums (TV, LCD, LED, Internet, mobile phone (smart series), iPod, etc.) at best. Media gadgets are developed to provide communication facilities effectively and efficiently. It becomes an intervening junction wherein both public and private authority serves their interests to make people follow them. Apart from the public and private dualism, media has matured as an institution to serve the peoples' interests over a period of time. In India, the urban population enjoys private media while the rural population has access to public media because the social, cultural and economic gaps restrict them from exploring options other than *Doordarshan* (DD—State-owned TV). However, print media is completely run by private players with government subsidy. In terms of circulation, Hindi and vernacular newspapers have outnumbered the English newspapers and have restricted its circulation to the urban population in India. As Vanita Kohili points out, "If it seems there is battle for news between TV and print it is incorrect. TV and print together account for large chunk of the money that news business generates, but the internet and mobile phones are fast emerging as serious contenders, at least in terms of volume."[13]

2.2 An Experiment with Media in India

The journey of the Indian media industry started under colonial vision with an aim to provide best media services to people. The media industry currently employs both "old" and "new" media resources under different organizational set-ups.

[13]Kohli (2006, p. 272).

Table 2.1 Top ten daily newspapers (2011–12)

S. No.	Publication	Language	Periodicity
1	Dainik Jagran	Hindi	Daily
2	Dainik Bhaskar	Hindi	Daily
3	Hindustan	Hindi	Daily
4	Malayala Manorama	Malayalam	Daily
5	Amar Ujala	Hindi	Daily
6	The Times of India	English	Daily
7	Lokmat	Marathi	Daily
8	Daily Thanthi	Tamil	Daily
9	Rajasthan Patrika	Hindi	Daily
10	Mathrubhumi	Malayalam	Daily

Source Media Research User's Council (Based on quarterly business)

The Ministry of Information and Broadcasting is looking after the daily requirements of the media industry and its future. The importance of media was incredible during the freedom struggle and since then, the role of media has unfolded many visions that have shaped the nation in order to make it self-reliant in this area. In this regard, print media was the pioneer in term of services provided to the people and in shaping the future of India in various contexts.

2.2.1 Tracing the Evolution of Print Media

In India, print media was the first medium of media practices which resumed its services under the British rule, despite facing a complete ban on printing newspapers on several occasions. The journey of print media in India started with the publication of the *Bengal Gazette* in 1779 by James Augustus Hickey and thereafter many newspapers were published from different location with different ideologies. Currently, there are more than 82,000 registered newspapers being published in India.[14]

The popularity of Hindi newspapers is greater as compared to other vernacular newspapers and English newspapers in India. After Hindi, Malayalam occupies the fourth and tenth positions while English, Tamil and Marathi newspapers are positioned sixth, seventh and eight. This shows a complete picture of the Indian print media industry and their proportional business shares (Table 2.1). Per Stahlberg points out that "the Indian press is generally categorised in broadly two sections—the English press and the vernacular press. However, it is also visualized as National and Regional press and overall the English press has a higher status

[14]For details report, see, FICCI-KPMG Report 2012.

than the vernacular press and consequently, the National press has a higher status than the Regional press."[15]

Since the expansion of the market for newspapers, the newspaper industry has broadened their base in order to expand their horizons to reach their readers through local editions. Almost all newspapers have opened their local editions under feasible market conditions. Robin Jeffrey described this phase as a "revolution," when the circulation of newspapers went up manifold, particularly between 1976 to the late 1990s. Jeffrey tries to explain the way in which this "revolution" takes place and, in doing so, acknowledged the increasing literacy rate and improved technology behind the scenes, which paved the way for tremendous increase in newspaper circulation.[16] Per Stahlberg identifies three reasons behind the expansion of the press. One is literacy, second is technology and the third is the ownership of media, which is shaping and reshaping the press in India.[17] These three factors happened together in India, resulting in the increase in the demand for newspapers over a period of time. Vir Sanghvi points out two reasons that enhanced the Indian newspaper business—the first is magazine revolution, and the second is the policy of liberalization.[18] Of course, growing literacy and the curiosity regarding the world and its affairs also played an important role in the increase in the number of people reading the newspaper. The popularity and curiosity to receive more news about politics becomes quite popular across the various sections of society, leading to the success of the newspaper industry in India. Meanwhile, politics comprises a major constituent of news. The reading of political news became popular after the emergency imposed by Indira Gandhi in 1975, when people wanted to know more about what was going on across spaces.[19] Robin Jeffrey points out that after the emergency was revoked and Indira Gandhi was defeated in the subsequent elections in 1977, newspaper circulation started to rise rapidly.[20] Zoya Hasan argues that the political development as a result was the decentralization of politics and the emergence of regional political voices that were noted in some parts of northern India apart from southern India, where the regional parties were already present. Gradually, over time, the various States took centre stage in Indian politics, parallel to Delhi centric news.[21] The changes in political activities were noted in states variedly and it provided an opportunity to the press to reach the various states, and the Hindi press gradually established itself over a period of time.

However, market competition amongst newspapers started to reach the reader's doors at the same time as the growing hunger for information was seen as being one of the reasons for increased consumption of newspapers in India. For this reason,

---

[15]Stahlberg (2013, p. 31).

[16]Ibid.: 53.

[17]Ibid.: 54.

[18]Sanghvi (2006).

[19]Stahlberg (2013, p. 53).

[20]Ibid.: 62.

[21]Ibid.: 63.

Table 2.2 Top ten hindi and english newspaper (2011–12)

S. No.	Hindi	English
1	Dainik Jagran	The Times of India
2	Dainik Bhaskar	Hindustan Times
3	Hindustan	The Hindu
4	Amar Ujala	The Telegraph
5	Rajasthan Patrika	Deccan Chronicle
6	Punjab Kesari	DNA
7	Navbharat Times	The Economic Times
8	Prabhat Khabar	Mumbai Mirror
9	Patrika	The New Indian Express
10	Nai Dunia	The Tribune

Source Media Research User's Council (Based on quarterly circulation)

every prominent newspaper started local editions to cater to peoples' need. These local editions have helped almost all popular newspapers to expand their horizon across the country. The circulation of newspapers has increased over a period of time and the following newspapers have established their credibility amongst readers (Table 2.2).

The experiment with print media, which started more than two centuries ago, is still continuing with greater emphasis on providing authentic news at reasonable prices to the readers with greater efficiency. The growth of newspapers in India is decipherable from Table 2.3. The gross circulation crossed more than 320 million while in terms of registered newspapers; Uttar Pradesh and Delhi emerged as the largest circulation centres in India.

In 2011, more than 4500 newspapers were registered, which shows people's faith in print media in India. Amongst the English newspapers, *The Times of India* emerged as the frontrunner in terms of circulation with 12 editions. *Eenadu* emerged as the frontrunner amongst vernacular newspapers with 19 editions and the maximum circulation in Hyderabad. *Anand Bazar Patrika*, another vernacular newspaper, had a circulation of more than 1.2 million copies in Kolkata alone. Uttar Pradesh became the pioneer for the expansion of the press, while Allahabad, Kanpur, Varanasi, Lucknow, Gorakhpur, Meerut and Agra became the nodal centres in the states where regional editions of many newspapers are circulated.

2.2.2 Mapping the Growth of Radio

Radio service started in early 1920s and the first programme was broadcast in 1923 from the Radio club of Bombay. In India, radio broadcasting services started with the help of two privately owned transmitters—one at Bombay (Mumbai) and another at Calcutta (Kolkata). The Government of India took custody of both transmitters in 1930 and resumed services under the name of Indian Broadcasting

Table 2.3 Newspaper circulation figures

S. No.	Circulation			Figures
1.	Total number of newspapers registered as on 31st March 2010–2011			82,222
2.	Number of new newspapers registered in 2011			4853
3.	Total number of circulations			329,204,841
4.	Largest number of newspapers registered in any Indian language (Hindi)			32,793
5.	Second largest number of newspaper registered (English)			11,478
6.	State with largest number of newspapers (Uttar Pradesh)			13,065
7.	State with second largest number of newspapers (Delhi)			10,606
8.	Largest number circulated daily:			
	Newspaper	Language	Place of publication	
	Eenadu	Telugu	Hyderabad	16,74,305
	The Hindu	English	Chennai	14,82,658
	Ananda Bazar Patrika	Bengali	Kolkata	12,08,682
9.	The largest circulated multi-edition daily:			
	Newspaper		Number of editions	
	The Times of India		12	37,46,929
	Eenadu		19	29,55,300
10.	Largest circulated periodical is The Hindu weekly			13,48,160

Source Media Research User's Council (Based on Quarterly business)

Service. Later, the name was changed again to All India Radio and since 1957, it has been known as *Akashvani*.[22] At the time of independence, there were six radio stations and 18 transmitters, which covered 11% of the population and 2.5% area of the country. By the year 2010, there were 232 AIR (All India Radio) stations and 374 transmitters, which provided coverage to 99.16% population and 91.82% area of the country.[23]

Akshay Rout argues, "Radio-cheap and portable, friendly and intimate-gets through the illiterate, the less formally educated, and the lower socio-economic strata, all of whom constitute the priority audience in public health, basic education, and rural development."[24] He further infers that during the green revolution, Radio played an important role and made *Grow More Food* campaign such a success that the hybrid rice produced was known as "radio rice." With time, radio became a popular media gadget for listening to songs, news, sports commentary and theater programmes including imparting agricultural news to farmers for taking better care of crops. It is commonly believed that the use of radio is the most widespread in rural villages and in the remotest locations where physical accessibility is very

[22]See India (2012, p. 755).

[23]See India (2012, p. 757).

[24]Rout (2006, p. 138).

challenging, while FM radio is quite popular among urban dwellers, especially amongst the youth. The use of radio is inevitable across the coastal localities for various purposes such as to know about the weather conditions of the sea or to know the ongoing market price of fish.

The contours of radio changed over time as it used FM radio, digital broadcast, satellite radio, web radio, etc. Digitalization, of course, enhances the quality of the radio services including storing, retrieving and sorting of news and other programmes. The entertainment—education component in a programme gives an edge to radio over other kinds of media. In fact, radio needs less knowledge of proficiency in terms of handling it. The spatial expansion of radio programmes has increased over time and radio prepares programmes for all sections of society vividly to depict the cultural vibrations.

Since the past eighty years, radio has been serving the nation very effectively and efficiently even during crisis periods of the Freedom Struggle, Indo-China war, Green Revolution, liberalization, the second nuclear test, Kargil war and others. AIR (All India Radio) is a state-owned radio transmitter and is supposed to reflect the views of the state on various issues. To deliver news in a responsible and accountable manner is one of the basic intentions of the AIR. "The NSD (News Service Division) of AIR produces 112 bulletins in 17 languages that are broadcast from Delhi and relayed by several AIR stations."[25]

Since 2001, FM radio has been allowed to broadcast news with certain restrictions and the "FM-ization" of news has been shaping and reshaping Indian audiences in various ways across urban localities.[26] FM played an important role in setting a good precedent for the dissemination of news at a fast pace in India. The expansion of FM radio undermined the main radio programmes and indicated a problem wherein AIR's news broadcast was losing control. Rout further explains through various examples that some time in the future; we might have to listen to private players in the form of *Dare-Bare News, Khatta-Mittha News* (sour-sweet news), *Masala News* (spicy news) and so on. Furthermore, the private FM could impinge on AIR's traditional holdings. However, on FM radio policy, Amit Mitra in his report recommended that private players should be allowed as long as they abide by AIR's code of ethics.[27]

Marshall McLuhan describes radio as "the tribal drum" and "hot"[28] medium where he finds massive experience of electronic implosion. Describing McLuhan's experience, Kanchan Kumar says

> Radio for him [McLuhan] is an earthy medium that has native power to involve people in one another and resonates a web of kinship. He [McLuhan] argued that radio revives the primordial 'acoustic space' of oral societies and thus creates an environment of simultaneous experiences and sensory unification…As a medium for news and reports, radio has

[25]Kumar (2006, p. 156).
[26]Rout (2006, p. 145).
[27]Ibid.: 148.
[28]McLuhan (1964/2001, p. 324).

the advantage of adaptability for localized coverage of news, events, and community-based programmes. It can easily score over both TV and newspapers as news dispenser as it engages only the aural facilities. Radio can be heard while doing other things such as driving, travelling, cooking and hence is the most accessible medium for catching up with the events of the day without having to suspend routine activities.[29]

The utility of radio in a developing society is inevitable in many ways for economic, cultural, social and geographical knowledge.

2.2.3 *Emergence and Growth of Television*

The Television saga started with the installation of terrestrial television in New Delhi on September 15, 1959 to transmit educational and developmental programmes on an experimental basis of half-an-hour programming. The commencement of regular television service was operational as part of AIR, which later commenced in Delhi (1965), Mumbai (1972), Kolkata (1975) and Chennai (1975). *Doordarshan* (DD) was established on September 15, 1976 and was supposed to telecast the official version of the State.

TV commenced its activities in black and white colour combination and over time it progressed into colour. The nineteenth Asian Games held in Delhi revolutionized the concept of television and its aesthetic pleasures. After that spatial expansion, TV programmes were broadened and transmitters were installed to broaden the services. The liberalization policy in 1991 helped the expansion of TV and allowed private players to run private TV channels in India. In India, around 828 private TV channels have been given permission to do their business so far, out of which around 404 of them are running news programmes of myriad forms.[30] These private players have revolutionized the TV industry and the content for programming for example in case of news, soap operas, music, advertising, sports, infotainment, movies and devotional programmes.

The number of television sets has increased every year by an average growth rate of more than five percent while cable and satellite connections have increased by more than ten percent. The process of digitalization enhanced the picture quality and other administrative purposes and this process should be completed by the end of 2014. The process of digitalization is complete in Delhi and Mumbai while it is in progress in Chennai and Kolkata and other metro cities. This process of digitalization should be complete in India by the end of March 31, 2015. It is evident from Table 2.4 that the number of television sets is growing across households and the popularity of cable satellite is notable across households as well.

---

[29]Kumar (2006, pp. 156–57).

[30]See for a detail lists prepared by Ministry of Information and Broadcasting, Government of India, New Delhi, http://www.mib.nic.in/WriteReadData/documents/status_220313.pdf, accessed, March 13, 2013.

Table 2.4 Number of TV, non-cable & satellite, cable & satellite, and digital (all India household figures)

Households/figures	2009 (in Million)	2010 (in Million)	2011 (in Million)	2012 (in Million)	2013 (in Million)
TV owning	123	134	142	148	153
Non-cable & satellite	33	30	26	21	13
Cable & satellite	90	103	116	126	140
Digital	15	20	26	42	59

Source TAM Media Research, India

Table 2.5 Urban and rural media consumption

TAM annual universe	Urban (in Million)				Rural (in Million)			
	2010	2011	2012	2013	2010	2011	2012	2013
TV owning households	64	68	69	71	70	73	79	82
Cable & satellite households	54	60	63	66	49	56	64	74
Digital households (DTH subscribers)	6	9	16	25	14	17	26	34

Source TAM Media Research, India

Daya Kishan Thussu argues that in the late 1990s and early 2000s, India's news television sector demonstrated extraordinary growth in the number of dedicated news channels—from 1 in 1998 to nearly 40 in 2007. Most of them were national, but some of them international in reach, while some delivered regional imprints.[31] So far, the number of private news channels in India is 404 and in future there are chances of the number of number of private channels increasing manifolds. The Doordarshan network consists of 64 Doordarshan Kendras/Production Centres, 24 Regional News Units, 126 Doordarshan Maintenance Centres, 202 High Power transmitters, 828 Low power transmitters, 18 Transponders, 30 Channels and DTH Service stations. While following five channels are national channels viz. DD1, DD News, DD Bharati, DD Sports and DD Urdu.[32]

The parity between rural and urban India in terms of having a television set has reduced in the past few years but as per the population index, the gap between the rural and urban population remains wide (Table 2.5). The number of television sets is set to go up to 200 million by the year 2017. The preference of cable, satellite and DTH connections over signal receivers has multiplied manifold in both rural and urban India. However, both public and private news channels have their own obligation to impart value added information based news or to propagate or

[31]Thussu (2007, pp. 96–97).

[32]See India (2012, pp. 774–75).

sensitize people to a certain extent wherein individual voices can be discernible over an issue. DD carries national responsibility and follows the government bytes on issues, while private channels promote commercial interests and try hard to maintain the neutrality and rationality in news content. Thus TV is a popular medium to impart spatial information in the form of "live" news. Rajiv Kumar elucidates, "DD is still struggling to find its real identity and is vacillating between being a department in the Ministry of Information and Broadcasting and trying to acquire independent status as a wing of the autonomous corporation of Prasar Bharti."[33]

However, both public and private TV channels become an eye of the country and cater to the requirements of the people in various ways. While DD has its own focus and runs stories which are in the interests of the country and keeps the national interest ahead of all other agendas; it portrays its content accordingly. Whereas, private channels have separate locus of interests and business is indeed their first preference and they often exploit the Indian cultural epitome to sell the identity of the nation.

2.2.4 The Genesis of Social Media in India

Social media, often called social networking sites, are gradually penetrating areas beyond the urban to the peripheral localities in India. Social media, which works through Internet connectivity, has resulted in urban people having an edge over the rural population in terms of usage. As Tim Jordon points out, "A growing body of theoretical and empirical work in all areas of the social science argues that sometime in the 1970s a fundamental shift in the nature of society was initiated. At its simplest, this is portrayed as the third great shift in human society, from agricultural to industrial and now to informational societies."[34] According to the Internet and Mobile Association of India (IAMAI),[35] around 105 million people were using the Internet by the end of 2012 and the number went up by June 2013 by around 111 million, out of which 89 million were considered active users while at the end of 2012, around 84 million were considered active users. At the end of 2012, around 64 million people were using social media while by mid-2013, the number was supposed to touch 66 million (Table 2.6). As Anuradha Parasar has pointed out, "Evidences support that since 1990 Internet usage can be characterized as the domestication of the Internet. By domestication, it means a shift from formal use to informal use that is using personal computers for household reasons."[36]

---

[33]Kumar (2006, p. 156).

[34]Jordon (1999, p. 145).

[35]IAMAI 2012 reports see, http://www.iamai.in/Upload/Research/31220132530202/Report-Social-Media%202012_67.pdf, accessed, February 12, 2014.

[36]Parasar (2006, p. 76).

Table 2.6 Urban India internet landscape

Claimed internet users	
105 MN (Dec 2012)	111 MN (Dec 2013)
Active internet users	
84 MN(Dec 2012)	89 MN (Dec 2013)
Social media internet users	
62 MN (Dec 2012)	66 MN (Dec 2013)
Note IMRB i-Cube 2012/Census 2011. All India estimates-June 2013	

Source IAMAI, 2012

In India, the notion of a "Global Society" is evident wherein the use of social media is growing at a fast rate in the urban centres and in the rural–urban fringes as well. As Ashwin Razdan points out, "One of the most popular features that resulted from the concept of web 2.0 is social networking. This gives an opportunity to interact with people who think alike. Online social networking has become a platform for people of all age groups to express themselves. It has led to the information of, what I would like to call, a 'Global Society'. A society where people from diverse backgrounds connect, irrespective of their ideologies, and eventually form a bond with each other. It has been noticed that most of the times the emotional involvement of the user is quite high. This has given birth to niche communities and websites with special interests and a separate set of user profile."[37] As per IAMAI, in the year 2012[38] around 80 million active users were using the Internet in urban India, wherein 72% (around 58 million) were using or accessing some form of social networking sites in India in various capacities. It is evident from Fig. 2.1, that out of 80 million active users of the Internet, 80% used it for email purposes, 72% for social networking sites, around 60% for entertainment purposes, 59% for gaming purposes, while 56% browsed for various reasons. Furthermore, around 43% users were using the Internet for utility purposes and interestingly around 41% users for chatting purposes.

The use of mobile phones for Internet purposes are widely observed today (IAMAI, 2012. p. 4), particularly the use of smartphones has gradually increased over time because smartphones allow easy access to the Internet. The craze of carrying smartphones among different sections of people has been growing for this reason.

As per IAMAI-2012, it is evident from Fig. 2.2, that the youth is increasingly using social networking sites and amongst the 80 million active Internet users, the profile of school going youth using social media (8–17 years old) is around 68%. The college going section (18–25 years old) comprises around 82% in terms of usage of social media and the highest proportion of social media usage was

[37]Razdan (2008, p. 83).

[38]IAMAI reports see, http://www.iamai.in/Upload/Research/31220132530202/Report-Social-Media%202012_67.pdf, accessed, February 12, 2014.

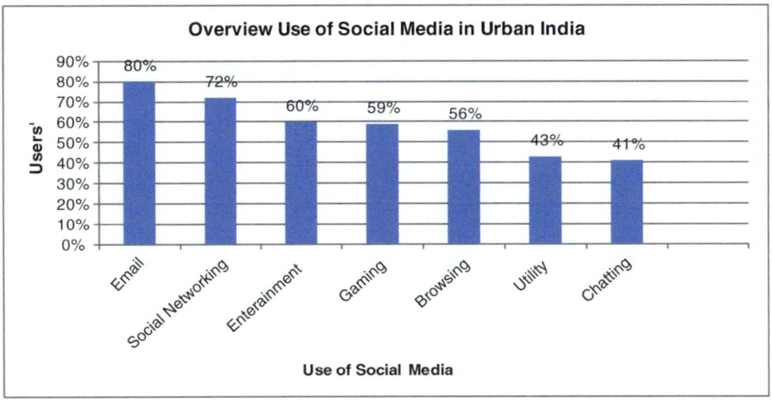

Fig. 2.1 Overview of use of social media in urban India. *Source* IAMAI, 2012

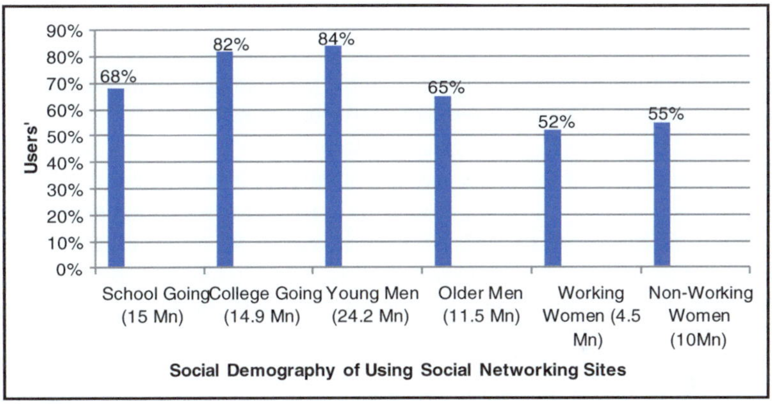

Fig. 2.2 Social networking profile by demographic segments (proportional). *Source* IAMAI, 2012

seen amongst young men (21–35 years old), about (84%). Working women (21–58 years old) and nonworking women (25–38 years old) also use social media with their percentage of usage being 52 and 55% respectively, while around 65% older men (36–58 years old) were observed using social media in India. It is interesting to note that the craze for social media is high amongst nonworking women as compared to working women. The young generation's interest in social media is indeed shaping the activities and users simultaneously.

The usage of social networking profile in towns is evident in Fig. 2.3. Across the top eight metro cities, about 35% users use social networking sites, while small towns with a population of less than 5 lakhs has about 35% people using social networking sites. Small metros and non-metro towns together comprise 30% of the population that uses social media. The penetration of social media in small towns

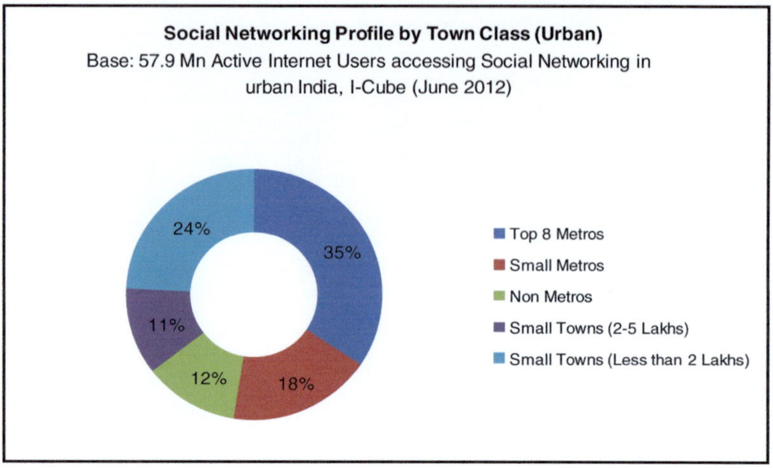

Fig. 2.3 Social networking profile of urban towns. *Source* IAMAI, 2012

with a population of less than 2 lakhs is notable while in non-metros and towns with a population of 2–5 lakhs is almost similar in usage patterns.

There are 23.6 million internet users from 35 cities in India, as is evident from Fig. 2.4. Around 77% of the users use mobile phones as a medium to access social networking sites, but out of these, around 33% use it for social networking as their main activity as compared to 32% who use it mainly for email usage. These users barely used the Internet for accessing online news, while 4% preferred to play online games and 10% used their phones for text chats. In other words, email and

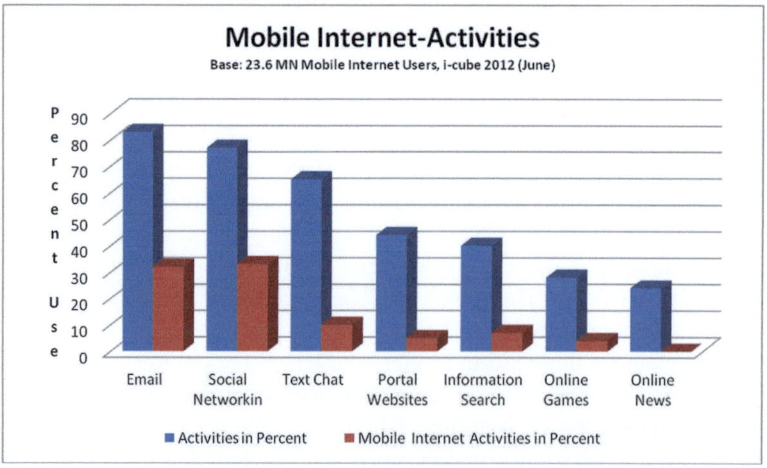

Fig. 2.4 Mobile internet-activities. *Source* IAMAI, 2012

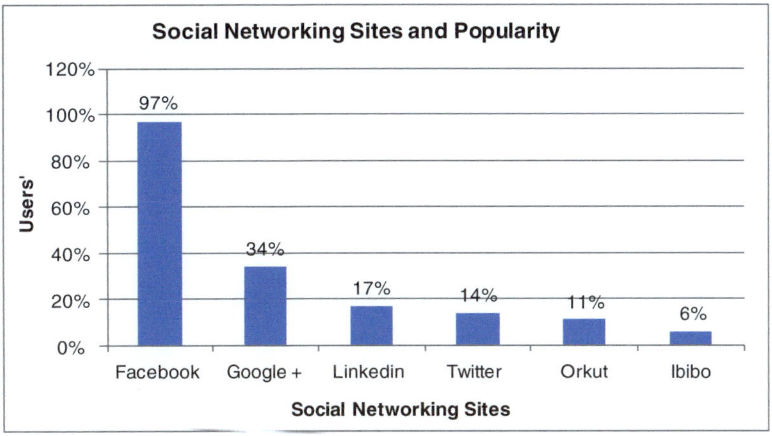

Fig. 2.5 Social networking sites and popularity. *Source* IAMAI, 2012 (The above data compiled from WAM (Web Audience Measurement system is known as WAM) under IMRB International. *Base* All India active internet users 51.33 MN/month from April–September 2012, 49.6 million users)

social networking are the major activities which people preferred to access through their mobile phones.

The popularity of social media is evident from Fig. 2.5, wherein around 97% people prefer sites such as Facebook, Google+, Linkedin, twitter, Orkut and Ibibo for business in India. It is interesting to note that the "professional networking site Linkedin has the highest proportion among females and above 25 years. In fact, Linkedin has the second highest average time spent only after Facebook."[39]

"The time spent by Indian netizens was 29.6 min on weekdays as compared to 28.8 min on weekend days while in other words, the amount of time spent is quite similar."[40] The time spent on weekdays is almost the same as the time spent on weekends, as is evident from Table 2.7.

The popularity of social media can be analyzed through the activities which the users follow in their day-to-day patterns on social networking sites. This is evident from Fig. 2.6, wherein around 73% of the users were using apps to connect with friends and share personal information. On Facebook, around 11% users were chatting or messaging, while 11% of the users were watching videos, 34% were looking for friends and 21% were accessing sites for looking at photos.

The popularity of social media in India is evident across urban locals, especially amongst the youth which includes school and college students. The use of social media is notable between working and nonworking women and in fact mobile technology, particularly smartphones or, touch screen mobiles which are

[39]See IAMAI report, 2012. p. 8.

[40]Ibid.: 8.

Table 2.7 Time spent on social media

Total minutes of usage *Base* All India active Internet Users-71 Mn, *source* WAM (September, 2012)	
On weekends	28.8 Min
On weekdays	29.6 Min

Source IAMAI, 2012

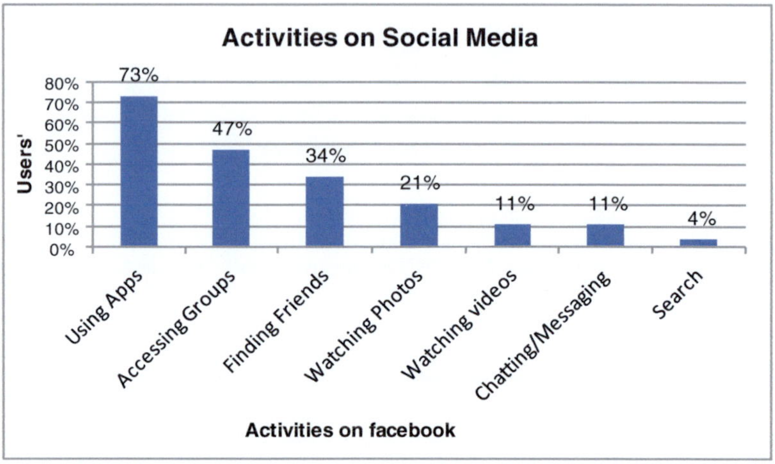

Fig. 2.6 Activities on social media. *Source* IAMAI, 2012

compatible with 2G and 3G technology have played an important role in revolutionizing social media in India. The role of social media and its impact on society is not distinctly clear but there is speculation that the impact of social networking sites is inevitable, especially since the Anna Hazare episode.[41] The Government of India has reiterated the need for proper mechanisms that can effectively check the wrong doings of social media in India. Recently in Muzaffarnagar (UP), where communal riots broke out during August–September 2013, people blamed social media because a contentious video was uploaded on YouTube by some miscreants, who were seen as being responsible for the spread of communal tension in the district. However, the district administration seized the video and controlled the ground situation after more than forty people from both communities were killed.

[41]Anna Hazare starts hunger strike for implementation of strong Lokpal to curb down the menace of corruption in the country however, he mobilize the peoples against the central government wherein role of the media was evident. The movement was successful at certain extent and aware the people to stand against corruption.

2.3 Negotiating Everyday Media in India

The role of media in everyday activities is inevitable. During the independence struggle, the role of media was to make people rise up against Colonial rule. In this, the role of print media was catalytic and mobilized people to participate in the agitation. Since then, print media has been serving the nation and its people at best with some changes, which includes news presentation. The only difference between then and today is that now it is considered as a part of business and earlier it was dedicated to the cause of uniting the people to fight against Colonial forces. Over time, the nature of media has multiplied in the form of "new" media (social media including Twitter) dimensions and has been overwhelmed by technological up-gradation (Gainous and Wagner 2014). Media influences space in several ways including social change as well.

Amongst visual display and "visual pleasure",[42] cinema was one of the initial mediums to get enlivened. There was scope for private players in the film industry and the growth of cinema halls was phenomenal after the independence in the country. Sometimes, cinema halls, or "theatre halls" used to show programmes apart from screening film shows. Henri Lefebvre reads the theater hall as being a "Third space."[43] Cinema halls have gradually transformed into multiplexes and are suited for "mall culture" , where multiplexes allow for the screening of multiple films at the same time. Across the metros, shopping malls have mushroomed where all kinds of facilities are available including places for shopping, eating and entertainment. The popularity of cinema halls changed over time and gradually made way for multiplex theaters. The role of media thus, is inevitable where the role of the state is restricted and reduced to merely issuing licenses and certificates.

Media became a phenomenal catalyst in terms of decentralizing their features on social networking sites where people can login and express their views without any technical prejudice. These sites encourage people to share their views through the available applications and by providing a platform for discussion. Over time, these media platforms have become important in forming a consensus over several issues. The fundamental characteristics of these media outlets is the facility to share attitudes, which has made them unique amongst the available media options where people share their views very openly and clearly without any prejudice. How do these media outlets function as an expression of popular thought? How do they shape and reshape the social identity of a space/place in various ways? How do these media outlets inspire an individual to do something for society? How do media help in building a consensus or propagate news to their followers? Why do people share their views freely over these mediums without fear of any repercussions? In 2006, a Bollywood movie *Rang De Basanti* directed by Rakeysh Omprakash Mehra, starring Aamir Khan, showed students protesting at India Gate (a war memorial located in New Delhi) against the government. In the movie, India

[42]Mulvey (2001, reprint 2005, pp. 393–404).

[43]Lefebvre (1991, p. 188).

gate is a symbol, denoting the government's position, and thus, the protest at India Gate becomes symbolic of protesting against the State's decision over an issue.

Ever since the success of the film, the India Gate monument has become a popular space for protesters, where ordinary citizens use the space to stage protests against government apathy towards their demands. Over the past few years several candle marches have been organized at India Gate to protest against the government. How do spaces become symbolic representation of a government? Like seen with the film *Rang De Basanti*, what galvanizes the people's attitude to make and locate a space where they can protest and register their voice of dissent firmly?

How does the selection of space take place for the purpose of protest? Why is a potential space required? Moreover, while news and other developments about the protests were circulated by audiovisual media, the topic was discussed extensively over social networking sites and visual media (satellite news channels) very intelligently and without forcing the people to come together at a physical location. Instead the protest took place in a fragmented manner. Given this, strategically, media wishes to sustain issues for a prolonged period of time as it increases their popularity and circulation. They even resorted to making the protest lively and engaging through exclusive coverage and by pushing emotional outbreaks from those involved.

During early 2011, another massive gathering of people stood for more than a week at Jantar Mantar, near India Gate, under the leadership of social worker Anna Hazare to fight against corruption and the non-implementation of the provisions made in Jan Lokpal bill.[44] The issue became contentious between the government, people, and other political parties during this period. The issue still exists, but not with as much fervour as in its initial phase was. During this protest, Jantar Mantar became the space/site for contestation and since then, it has taken a kind of centre stage, where many protests have been held. But the Anna-led protest gave a new identity to this space to raise one's voice against the government. The participation of the media was inevitable in this protest and some sections even accused the media of aggravating this movement against the government. While Jantar Mantar existed as a space for protest prior to the Anna movement, the media coverage of the movement gave a new lease of life to this space.

In December 2012, once again a massive crowd marched towards India Gate to lodge their protest against the government. This protest was organized for the rape victim Nirbhaya, as a show of solidarity towards her suffering and to support her cause. The protesters also demanded stronger laws against rape. Once again, the media, especially social networking sites, played a key role by providing a platform to people to share their common feelings of anguish against the crime. Over a week long agitation, the protesters were forced by the police to register their protest at Jantar Mantar, near India Gate, where entry of people was completely stopped for security reasons. Thus, both India Gate and Jantar Mantar emerged as public spaces where irrespective of their numbers, individuals or groups of people can lodge their

[44]A blueprint of the proposed Lokpal bill had been prepared by team Anna members.

protest against any authority. Both new and old forms of media extended their unconditional support to this youth agitation against the government and became mediums that allowed for the venting of feelings against the government.

2.4 The Political Economy of Indian Media

The political economy is a complex interrelationship between culture, economy, social movements and state (Granham 2001; Herman and Chomsky 2001). In India, media established as industry, parallel to other industry, is developing programmes in order to do business to benefit other as well.

This section is aimed towards understanding the basic strands of Indian media and whether they function to serve public interests or follow the path of corporatization. In order to do so, it is notable to examine the points made by Bennett (2007) in context to USA to support the stages when public media transforms into corporate media. However, what W. Lance Bennett suggested for USA is applicable in Indian situation as well. This includes the ways in which Indian media has pushed itself towards the corporatization of its business and the ways in which the media provides different news spaces for different purposes.

The various facets of Indian media flourished post-liberalization, when technology increased precision and enhanced the speed of media penetration across peripheral localities. The growth of media supports the gradual social change in society and acknowledges the freedom of speech to some extent while still much more needs to be done for it to become the voice of the poor, the marginalized or that of minorities. These voices of the marginalized sections of society surface as various programmes across media on screen (TV, cinema, mobile, iPod, laptop, web page, blogs, etc.) and in print (newspapers, magazine, etc.). Just like any other industry, media emerges as an industry, even though the media industry existed in the form of print, radio, television and cinema prior to the liberalization policy. Since liberalization, the media industry was consolidated at various levels, particularly as an investment and has expanded both vertically and horizontally so far. The shape and size of the media industry has changed over time in India and every media organization has sharpened their programmes in terms of quantity and quality, to lure business towards them. How Indian media grow their market share? How do both media and market work for their mutual interests? As per the FICCI-KPMG Report 2012, the Indian Media and entertainment industry grew from INR 652 billion in 2010 to INR 728 billion in 2011, registering an overall growth of 12%. The sector is projected to grow at a healthy CAGR of 14.9% to reach INR 1457 billion by 2016 (TAM Research, 2012).[45]

There was a time when media was working under the tight surveillance of regulatory authorities and was completely functioning under the public domain

[45]See for detail, www.tamindia.com, accessed, February 12, 2013.

where the government played a dominant role and media was supposed to follow the licence rules formulated by the government. Since liberalization, global businesses started courting Indian markets. As a result, today there are more than 800 private satellite news channels functioning at various capacities in India.[46] In India, the growth of media has been shaped by private media houses wherein the role of Rupert Murdoch was incredible when the Star group revolutionized the TV industry and was considered the pioneer in infotainment, reality shows and news. Daya Kishan Thussu points that "As in many other countries, the greatest contributor to the trend towards infotainment in India has been Murdoch, whose pan-Asian network Star, launched in 1991, pioneered satellite television in Asia, transforming TV news and entertainment." He further argues that "Murdoch was responsible for introducing the first music television channel in India (channel V); the first 24/7 news network (Star News); the first successful adaption of an international game show *Kaun Banega Karorepati* (who wants to be a millionaire?) and first reality TV series (Lakme Fashion Show)."[47]

Chandan Mitra, a journalist cum politician and parliamentarian from BJP, argues that Indian news media is in a churning process and has changed overwhelmingly over the period of a decade in such a way that it is beyond recognition in present times. How has Indian media (both print and TV) made strategic shifts from the traditional to the "tabloidization of mainstream media" to make news profitable.[48] The growing practices of tabloidization have started to affect our daily cultural activities. Daya Kishan Thussu observes that media has upgraded their programmes in both style and content and has acquired a professional approach and presents issues very dynamically. Further, he illustrates the manner in which the news channels push political satire as a regular feature *The Great Indian Tamasha* and *Double Take* on NDTV 24*7; *Gustakhi Maaf* on NDTV India and *Pol Khol* on Star News are examples of such programmes. On the other hand, news contents have been declining across channels.[49]

W. Lance Bennett has discussed the five effects of the concentration of media ownership on news content, as discussed by media scholars in the past and these are[50]

1. Dominance by fewer players in local and regional market distorts advertising rates, forcing small, independent outlets to quit, sell out or change their formats resulting in less diversity in music, news and minority affairs programming.
2. The pressure of corporate self-promotion means less critical coverage of the media industry in general and parent companies in particular.

---

[46]See for detail, www.tamindia.com, accessed, February 12, 2013.

[47]Thussu (2007, pp. 593–611).

[48]Mitra (2006, p. 53).

[49]Thussu (2007, pp. 103–4).

[50]Bennett (2007, p. 237).

3. News content shifts to infotainment formats due to the entertainment focus of owners and the economic efficiencies of soft news, "reality programming," and human-interest features.
4. News is regarded less as a public service commitment or a prestige builder for the parent company, and it becomes just another product line in the race for profits.
5. Innovation in packaging and branding disguises declining information diversity and content distinctiveness.

This research is keen to find out whether these arguments apply to the Indian media industry in which the role of private players is inevitable or not and whether or not they, "explore each of these potential impacts of shrinking participation of corporate ownership of the national information system."[51] The following information trends have been discussed by W. Lance Bennett at length.[52]

"Information Trend 1: Less Public Service and Alternative Programming"

In India, cable media diversified their business into various sections which ranges from news, travel, cuisine, shopping, religious programming, entertainment and providing "infomercials" (which provide information on lifestyle products). In India, different media organizations have been compelled to think about their future prospects with respect to both existence and profit. In the digital age, except print media, all other media information content increased and competition among media houses shaped the media business in the country. Therefore, as a consequence, bigger media organizations thrived while smaller ones would lag behind in the lists at every level of business.

In India, there are more than 800 private satellite TV news channels[53] which are permitted to transmit their programmes. But, only a few of them have been popular across sections of people. For Hindi news, apart from regional and vernacular channels, there are only a few which have gained popularity. This is perhaps because they have expanded their business in such a manner that no one can grow parallel to them. For example, *Aaj Tak, ABP* News (earlier known as Star news), *NDTV* India, *IBN-7* and *Zee News* are the prominent Hindi news channels. Amongst the English news channels, *NDTV 24*7,* Times Now and *CNN IBN* cover India and each of them poses as being the most popular in comparison to the other. Apart from the national scenario, there are a few channels which dominate at the regional or vernacular level and cater to people's aspirations, along with fulfilling their own motives of making profit through various means including advertisements.

According to Registrar of Newspapers of India that there are more than 82,000 newspapers registered in India but only few of them are able to maintain their business at a comfortable level. In both Hindi and English, there are only a few

---

[51]Ibid.: 237.

[52]Ibid.: 237–249.

[53]See Ministry of Information and Broadcasting for more details http://www.mib.nic.in/WriteReadData/documents/status_220313.pdf, accessed, February 11, 2014.

newspapers which are popular among readers in the country. Amongst English newspapers, *The Times of India, The Hindustan Times, The Hindu, The Telegraph* have gained great credibility, made a respectable position amongst readers, while amongst Hindi newspapers, *Dainik Jagran, Dainik Bhaskar, Hindustan, Amar Ujala*, are currently doing respectable business in India.

Every day media organizations are experimenting with new technology, designs, ideas and concepts, to devise strategies to expand their business. E-news is one of the experiments in this direction to lure more readers since mobile technology has become sharper and more accessible. Television news channels also provide Internet news (e-news) in order to cater to those who miss listening to live news. They can now access the same news content by a click of a button. Every media organization has its own web portal in order to keep ahead in the competitive market.

Hence, technological developments have made it possible to view news on the mobile screen, to know the cricket score, to share the feelings and anguish over an issue—all through a click of a button. The page 3 or colour page across newspapers was designed in a way to lure the young readers towards the newspapers. The content of the colour page ranges from films, sports, exotic locales. The idea behind this was to make newspapers popular amongst the youth who gaze at the featured pictures without necessarily going through the featured information. The small screen (TV) has become popular amongst Bollywood producers and directors to launch or promote their film along with actor/actress and other supporting members to reach the audience where earlier such accessibility was earlier not preferred from a business point of view. Most of the news channels and newspapers have their own strategies to increase their readership and viewership in order to make profits.

Across the Hindi news channels, the ongoing popular Hindi serials, reality shows, cricket matches, cuisines, fashion trends are discussed at length for promotional purposes and these help captivate the viewers. For example, *Aaj Tak* gives extra mileage to cricket even during news hour while both *Zee News* and *ABP* news give Hindi serials and reality programmes more coverage apart from their news schedule, sometimes even including such news in the news hour. Parallel to Hindi news channels, English news channels feature technological gadgets and vehicles in their programmes apart from news hour, giving them more emphasis. *NDTV 24*7*, for example, organizes "Greenathon," a marathon run to create awareness regarding the environment and to popularize eco-friendly lamps. The 2012 Greenathon campaign actively used the services of popular Bollywood actors and actresses like Amitabh Bachchan, Shah Rukh Khan, Priyanka Chopra, Milind Soman, Rahul Bose and others along with cricketers such as Sachin Tendulkar; to make the programme engaging, attract more viewers, and disseminate the real message among the viewers. Being a part of the promotions for Greenathon, Bollywood Film actor, Milind Soman[54] ran a 1500 km long marathon from New Delhi to Mumbai to create awareness about the environment and said, "The world can be

[54]See for detail, interview www.ndtv.com/campaign/milindsoman, accessed, April 20, 2012.

happier and healthier if each person does his bit." In one of the campaigns on *NDTV 24*7*, "Marks for Sports" (2013), where the emphasis was on why sports is necessary to keep a person fit. The campaign was supported by sportsmen and Bollywood personalities. In his promotional speech for the campaign, Bollywood actor Ranbir Kapoor[55] said, "The country needs to take sports more seriously and not just as a hobby. As part of *Marks for Sports*, I hope to strive to do just that." Sportsmen joined the campaign and extended their support. Baichung Bhutia[56] showed his solidarity with the campaign and requested parents to come forward and said, "We are focusing on the kids for *Marks for Sports* in schools but it is the parents who really need to be involved in sports and fitness. I think the fittest contest will get parents involved as well." *Times Now* ran a campaign in association with *The Times of India*, "I Lead India-I will be the Change" (2013),[57] where through an advertisements, request made to the youth of the country, the campaign aimed towards encouraging the youth to make a change in themselves instead of blaming the government and take the nation ahead. *CNN IBN* launched a campaign in association with Penguin Books Summer Reads (2013),[58] in order to promote reading habits among students and youth during summer holidays.

Such campaigns are a part of the idea to promote social causes in society. Of course, the campaign is designed in such a manner that it attracts business and keeps the news industry profit oriented. These programmes (campaigns) also reduce the actual time spent in featuring news. One can see how in several ways, Indian media is diverging towards a point where news content is considered less public service and alternative programme.

"Information Trend 2: Internal Corporate News Censorship and Self-Advertising as News"

The media industry is considered as one of the fastest growing businesses in the country. Like other industries, the media industry set up their business on a model that paralleled the one followed by corporates and established the news business. The *Star group, Zee group, NDTV group, Times group, Aaj Tak* group, *CNN IBN, Sahara group* established their credentials in the news market by making news as the saleable product and packaged it accordingly.

Many news anchors even crossed the limit of what they were supposed to say and did so deliberately as part of reading the news. During prime time news, anchors in their news discussion make deliberately provocative statements or even talked out of context to validate their concerns regarding the issues. The competition amongst news channels is usually centred on who telecasts the concerned

---

[55]Bollywood film actor and gave so many hits films, see for detail, interview, www.ndtv.com/campaign/rabirkapur, accessed, July 13, 2012.

[56]Biachung Bhutia was leading football player and he represented India at both national and international tournaments many times. See for detail, interview, www.ndtv.com/campaign, accessed, April 6, 2013.

[57]See for detail, www.timesnow.com/ileadindia, accessed, March 16, 2014.

[58]See for detail, www.cnnibn.com/summerreading, accessed, June 13, 2013.

news first. The war amongst them to claim that they were "first" is quite dominant. During the protest march and tabloidization of the rape case in Delhi (Dec. 2012), Indian media came to a consensus that they will not reveal the actual name of the victim and hide her real identity and called her *Nirbhaya* (without fear) instead. However, the Delhi gang rape remained a top story for more than a week across media channels and newspapers for different reasons, whether to report the agitation or to discuss the action taken by the government or to mobilize people to stand up against the gruesome incident.

Similarly, during Anna Hazare's movement, the media had an opportunity to propagate the movement and take it to people's doorsteps and they made it very successful. The movement did not get adequate media attention in the second and third phase of their agitation as compared to the first phase. There was a lot of fuss amongst the people that due to the overt and covert support of the media to the movement; that it lasted for a week. On various news channels, the news anchors reiterated that the movement was a "mass movement" but it was actually restricted to a few hundred people. It is imperative to question why the media supported the Anna movement beyond a limit.

The manner in which news channels run some news episodes might be an indication of the corporatization of the news content, wherein they are focused on TRPs. Indeed TRPs is one of reference of its kind where particular news channels claim that they are ahead as compared to other channels. These claims are of course, to show that they are the leading channel. While TRPs are considered contentious, it still remains the most popular method out of all other available methods of measuring the popularity of channels among viewers.

Each news channel and newspaper has their own set of standards which they are supposed to follow to keep in line with the business and carries their distinct identity which is identifiable from their logo to the manner of news presentation. *NDTV* for example, never allows any hate speeches or warmongering speeches against Pakistan in their news hour and instead covers stories to pacify bilateral relations between the two neighbouring nations. *Times Now* does not give Kashmiri separatists a space in their news hour. Instead, they cover the highlights of the Indian perspectives over the issue. *The Times of India* launched "Aman ki Asha" programme[59] to foster the peace process between India and Pakistan.

Gradually, the Indian media industry became known by a few dominant names and they replaced the marginal and small media agencies. They developed a separate identity in the news industry and became quite popular amongst viewers. Therefore, we can visualize how Indian media is moving ahead to strengthen "internal corporate news censorship and self-advertising as news."

"**Information Trend 3: More infotainment and Soft News**"

A trend of infotainment has been recognized at a very primary level across media in various capacities, by giving more space to "soft news," to attract more viewers

[59]See for detail, www.thetimesofindia.com/amankiasha, accessed, May 13, 2013.

irrespective of age. W. Lance Bennett observed that television journalism has become "an adjunct of marketing and thus must be more entertaining and more adjusted to the fantasy world of drama and fiction."[60] The entertainment formula in the news is only for making news more viewer-friendly so they can consume it easily.

The reality shows are packaged in such a way that the viewer's senses are stunned. Most news channels show film songs to sensationalize news content. This is one way in which the anchors compare two episodes side by side and reiterate their stand over an issue. There is a growing tradition of telecasting soft news either in the beginning or at the end of news; to make news more attractive. Soft news ranges from entertainment to sports to infotainment. It also comprises commentary, art, lifestyle, travel and other such topics.

While the media industry is using soft news for such purposes, newspapers are experimenting with the content carried in the coloured pages, frequently switching from one theme to another in order to sustain the readers' interests. The photographs of college going students and other socialites including Bollywood actors remain main features of the colour pages. The way in which the photographs are featured was even considered the beginning of a new style and marked the beginning of a new model for the colour pages. Together, news channels and newspapers persevere to sustain viewers' attention by employing means such as infomercials, infotainment and tabloidization, to push their business forward.

"**Information Trend 4: Generic News**"

News channels have strategically developed a concept of TRPs to make their business competitive and to lure advertising agencies. In order to obtain higher TRP ratings, news channels and newspapers package news in such a manner that news loses its originality and is instead pitched as a saleable product. The technical advancements have also aided the media industry to transform the results through news content and other animation programmes.

In the competitive market, media houses employ readymade techniques to make news. For example, lab technicians add sound bytes over footage in order to prepare news. Alka Saxena discusses the glamor quotient involved in the news and argues, "it is assumed that the electronic media is glamorous so the backdrop has to be good and the set has to be visually appealing. Now a third aspect has been added to this- the anchors should be young and attractive; they should use professional make-up persons and hair-dressers so that they can keep up with this glamorous image. Today you find that news anchors are also mini-celebrities. They are mini-versions of film stars. But can all news be glamorous? I don't think so."[61]

[60]Bennett (2007, p. 240).

[61]Saxena (2006).

"Information Trend 5: Branding the News"

Alka Saxena further observes that the content policy of TV News channels is decided first by journalistic instincts and second by commercial interests.[62] *Doordarshan* (DD), being a state-owned channel, made its own identity and reflected the State's vision on various themes. Private media players inevitably carry different perspectives on issues. The English media has a completely different tone as compared to Hindi media in India.

Bennett contemplates the ways in which news is a product. He says:

> It [news] surely differs from cornflakes and underarm deodorant, yet it shares one important thing with most commercial products: it is marketed effectively through product images, which create the illusion of distinctiveness while often blurring the underlying realities of product quality and differences. One brand of cornflakes conveys an image of crunchiness, another exudes healthiness, and still another evokes an image of family togetherness at the breakfast table. They are all cornflakes. Similar to other products, the news is packaged. Packages convey images. They dress products up or down. They give off signals about the social status of the people who use particular products.[63]

So like other products, the media industry uses news as product and packages it in an aesthetic manner to lure audiences and to increase their TRP's in order to fetch more and more business in terms of getting advertisements. The Hindi news channels sometimes construct news in the studio and put all efforts to make it look like a real issue or actual content.

The nature of Indian media is growing in a "capitalist economic" structure as laid down by Bennett's five points of discussion. Since the liberalization of the economy in 1991, the Indian media industry has never looked back. The facets of Indian media are transforming every day in the sense of building a promising relationship between viewers and the media industry in materialistic ways. The small and marginal media offshoots have vanished or have been replaced by something bigger. The era of media houses in India has set the stage and is doing business primarily in the line of profit and not out of a sense of public service. However, the struggle between media and the state has transformed over time and both have set their own domain of preferences.

2.5 Conclusion

The use of media in our day-to-day life is growing everyday with new assumptions and perceptions where the importance of media has been acknowledged. Media become a representational space where representation of space and place is somehow experienced in spatial expression even without being there. Media emerged as an important source and experienced the spaces and places and

---

[62]Ibid.: 185.

[63]Bennett (2007, p. 247).

acknowledging for various purposes as well. It has become instrumental in shaping our everyday lives. However, the shape of Indian media is quite versatile where English and vernacular media including Hindi language media have their own domain of interests.

The post-liberalization business atmosphere in India shaped the media industry where more than 800 private channels have been functioning in different capacities. The privatization and globalization also shaped the Indian media industry and media gradually shifted from "public" to "private" wherein programmes are made to lure more business rather in the past public programmes are given preferences. Since liberalization policy of 1991, Indian media industry has opened the door for the regional programmes and has expanded its business as well. However, the content in the media programme were gradually shaped for greater coverage in order to lure market. Therefore, many experiments were conducted to made programmes better in terms of coverage's and economic viability. Meanwhile, the political content is indeed considered important constituents of media programming.

However, media was busy to cover UP Assembly election 2012 while four more states (Punjab, Goa, Manipur and Uttrakhand) where Assembly election was supposed to hold. Media knew the political importance of UP in the Indian political space which considered the UP Assembly election as prelude for 2014 Parliamentary election and as heartland of Indian politics and polity—those who win have chance to form government in the centre (Delhi) as well. During election campaign, media discussed all political permutation and combination with respect to UP whether it was caste politics or communal politics or most of Prime Ministerial candidates produced by the state or how it important for any political party whether regional or national party.

The next chapter is about to discuss in details the political positioning of UP in Indian politics that why it is known as heartland and at what extent important for political party who have political stake including national parties (Congress Party and BJP) in the state as well. The political battle between regional and national party through mediascapes are discussed with mundane political colour in the forthcoming chapter.

References

Adams PC (2009) Geographies of media and communication: a critical introduction. Wiley-Blackwell, Chichester

Appadurai A (1996) Modernity at large: cultural dimensions of globalisation. University of Minnesota Press, Minneapolis

Bennett WL (2007) News: the politics of illusion. Pearson and Longman, New York

Black J (1997) Maps and politics. The University of Chicago Press, Chicago

Crampton JW (2010) Mapping: a critical introduction to cartography and GIS. Wiley-Blackwell, Oxford

Gainous J, Kevin WM (2014) Tweeting the power: the social media revolution in American politics. Oxford University Press, New York

Garnham N (2001, reprint 2005/06) Contribution to a political economy of mass communication. In: Durham MG, Kellner DM (eds) Media and cultural studies. Blackwell, Oxford
Herman E, Chomsky N (2001, reprint 2005/06) A propaganda model. In: Durham MG, Kellner DM (eds) Media and cultural studies. Blackwell, Oxford
India (2012) Publication division, ministry of information & broadcasting. Government of India, New Delhi
Jeffrey R, Doron A (2012) Mobile-izing: democracy, organization and india's first "mass mobile phone" election. J Asian Stud 71(1):63–80
Jordon T (1999) Cyberpower: the culture and politics of cyberspace and the internet. Routledge, London, p 145
Kohli V (2006) The business of news. In: Sahay U (ed) Making news: handbook of the media in contemporary India. Oxford University Press, New Delhi, p 272
Kumar K (2006) Radio news: the Indian station. In: Sahay U (ed) Making news: handbook of the media in contemporary India. Oxford University Press, New Delhi
Kumar R (2006) The biggest public service TV news broadcasters: Doordarshan. In: Sahay U (ed) Making news: handbook of the media in contemporary India. Oxford University Press, New Delhi, p 156
Lefebvre H (1991) The production of spaces (Trans: Nicholson-Smith D). Blackwell, Oxford
Lundby K (2009) Mediatization: concept, change, Consequences. Peter Lang, New York
McLuhan M (1964) Understanding media: the extensions of man. Mc Graw Hill, New York
Mitra C (2006) Economics and politics of news: editors and owner's dilemma. In: Sahay U (ed) Making news: handbook of the media in contemporary India. Oxford University Press, New Delhi, p 53
Mulvey L (2001, reprint 2005) Visual pleasure and narrative cinema. In: Durham MG, Kellner DM (eds) Media and cultural studies. Blackwell, Oxford, pp 393–404
Parasar A (2006) Impact of internet on society. Aavishakar Publications, Jaipur, p 76
Razdan A (2008) New media: techniques and trends. Vitasta Publishing, New Delhi
Rout A (2006) AIR news: change beacons. In: Sahay U (ed) Making news: handbook of the media in contemporary India. Oxford University Press, New Delhi
Sanghvi V (2006) Making of newspaper. In: Sahay Uday (ed) Making news: handbook of the media in contemporary India. Oxford University Press, New Delhi, p 54
Saxena A (2006) What makes TV news. In: Sahay U (ed) Making news: handbook of the media in contemporary India. Oxford University Press, New Delhi, p 185
Stahlberg P (2013) Writing society through media: ethnography of a hindi daily. Rawat Publications, Jaipur
Thornaham S et al (2009) Media studies: a reader. Edinburgh University Press, Edinburg, p 497
Thussu DK (2007) News as entertainment; the rise of global infotainment. Sage Publications, Los Angeles

Chapter 3
Uttar Pradesh: The Heartland of Indian Politics and Polity

Abstract Prior to the independence and even after that UP emerged as "Heartland" of Indian politics for; geographical, social, cultural, political and linguistic reason. The making of heartland was strengthened by the number of seat allocated in the parliament or number of maximum PM produced or those political party rules in the state have greater chance to rule in the centre as well. Caste and communal politics made the state politically sensitive which have potential to polarize the society and politics beyond the state territory. Media tried to map the political glimpses with the help of eight political issues which considered contentious across the political parties during election in the state.

Keywords Heartland · Identity · Caste politics · Communal politics · Mediated

Uttar Pradesh (UP) is a politically dominant state and politicians generally consider that the path towards Delhi comes through Lucknow, the capital city of UP. If any political party wishes to rule the country that political party must have roots in the state and implicitly or explicitly should control the politics of UP. Only then will there be a possibility to rule over Delhi. The political contribution of UP is inevitable in national politics, either for coalition or for supporting the government from the outside.[1] Apart from political identity, UP has the maximum population and is known as the hub of "Hindustani culture" and Hindu-Muslim unity. The landscape of UP is not only about political activities, rather it is known for agricultural products, cultural taboos, cultural etiquette, communal politics and caste politics as well.

In the year 2012, five states went for Assembly elections, including UP but news about other State Assembly elections was proportionally less in comparison with the UP elections in the media. Media knew the political importance of the state and particularly in coalition politics when UPA II survived with the support of the

[1]Both SP and BSP had supported the Congress Party led UPA II from outside.

S. Moinuddin, *Mediascape and The State*, Springer Geography,
DOI 10.1007/978-3-319-51932-6_3

Image 3.1 Politicians dancing on public tune. *Source* TOI. 8.3.2012. p. 12

Samajwadi Party (SP) and Bahujan Samaj Party (BSP), the major regional political parties in the state. Why did the media show such interest in the politics of UP over other states in the country? The study of UP Assembly elections 2012 is pivotal for various reasons including the fact that it involves a political battle between national and regional political parties and was crucial in shaping and reshaping the national politics as well. The role of UP in national politics is known either for providing maximum number of Prime Ministers or for sending the maximum number of Parliamentarians. The UP election was considered a decisive election in various ways—it determined the future of UPA-II government, based on whether Mulayam Singh Yadav would withdraw support or BSP would extend support to the UPA-II, the Congress-led UPA-II government would fall or there would be ample chances for Parliamentary election after the UP Assembly election 2012. All these discussions shaped the UP elections as a decisive election in the media.

In a democratic setup, elections remain a political event, when people choose a representative, irrespective of political ideology. Hence, election is the only moment when politicians are supposed to dance to the tune of the people (Image 3.1). In the picture below, which was featured in *The Times of India*, one sees two people talking to each other and one of them is saying, "I know that nothing will change for us, but I must say that I do enjoy these item number competitions we get to see thanks to the polls."

During elections, the common people feel empowered to choose his/her candidate for good governance. The political landscape becomes vibrant when the election campaign starts in the states. This chapter discusses how media shaped UP as the "heartland" of Indian politics and polity. It also looks at the inherent caste politics, communal politics, centre–state politics and electoral politics that shape it into the "heartland" of Indian politics and polity.

3.1 Creating the Heartland

Kudaisya (2006) points out, "UP has came to decisively influence India's politics and to be constructed as a political 'heartland' through much of the twentieth century-both colonial and post colonial."[2] The construction of the political "heartland" started during the early years of the twentieth century and continued till the sixty seven years of independence and the political construction was evident in the UP Assembly elections in 2012 as well. UP can be perceived as heartland in multiple ways and explores five such distinct, though overlapping, constructions through its modern history: UP as a "colonial heartland," as a "nationalist heartland," as a "Hindu heartland," as a "Muslim heartland," and finally as a "post-colonial heartland."[3]

The colonial framework was constructed in order to establish British rule in the state particularly after the 1857 revolt. The 1857 revolt put pressure upon the British to mull over its administration and shape and plan for the future. As Kudaisya points out that

> Deeply impacted by the experience of the revolt, the new police created by the British had a new priority: defence of the realm, rather than prevention and investigation of ordinary crime. Its chief concern remained the maintenance of public order, rather than the protection of the common man and his property. The common man thus perceived the policeman largely as an agent of the *sarkar*, rather than as a guardian of law and justice.[4]

During the colonial period, the essential component of the *Pax Britannica* was the codification of criminal law and the establishment of a system of criminal justice.[5] Apart from this, the British made certain "riot plan" to control the crowds and the gathering of people when required earnestly.[6] The "colonial heartland" was an initial stage when British made elaborate plan to tackle the law and order situation to keep the state confined under colonial rule. The colonial *Pax Britannica* was maintained through "the internal security garrison of the army and the police. It has shown that the army played a vital role in the overall machinery that was created by the *Raj* for the maintenance of public order."[7] Kudaisya elaborates that the British experimented through "chaukidars" and the "mukhia" largely but these two institutions however, came to be eroded in the late nineteenth century from the point of view of the Raj and had largely become dysfunctional by the 1930s. "This forced colonial officials to depend mainly upon the influence of landlords and rural elites to maintain a semblance of order in rural society. In the context of UP the landlords were particularly important. They were regarded as the 'bulwark of the

---

[2]Kudaisya (2006, p. 6).

[3]Ibid.: 7.

[4]Ibid.: 86.

[5]Ibid.: 86.

[6]Ibid.: 87.

[7]Ibid.: 150.

Raj,' and the principal agents of authority in the countryside where no formal agencies of control existed."[8] They made all arrangements to control the local administration whenever the situation arose.

The sense of "nationalist heartland" shaped during the independence movement when "the civil disobedience campaign unfolded; the government was increasingly forced to depend upon extraordinary measure provided by the ordinances to restore its authority. Liberal hope of containing the campaign through ordinary law soon vanished; indeed, the law could no longer remain a neutral arena when dealing with the Congress. The government attempts to criminalize political prisoners were rendered ineffective, for those it designated 'criminals,' the public feted as heros! In such a scenario, continued use of criminal law could undermine respect for legal system itself. So, colonial officials instead acquired special powers through ordinances to contain the mass movement. Evolving a new punitive framework to deal with the political agitation, however, entailed certain costs. Trials and prosecutions against Congress volunteers were no longer seen as attempts to maintain law and order but as part of a larger, pernicious attempt to uphold the power of the Raj in a coercive manner."[9]

> Similarly, popular attitude of jail-going for political activities were altered. Before World War I, going to jail or even to the police station was regarded as an act of shame beyond the sensibilities of decent people, akin almost to the loss of caste for Hindus. This view was not just a middle-class perception, but also held in the countryside where village elders regarded inquisitorial police visits as a collective embarrassment. By the 1930s, these attitudes were changing. Nationalist discourse now represented jail-going in metaphors of pilgrimage, sacred duty, indeed sacrifice.[10]

The "nationalist heartland" indeed shaped the perception of individuals that the nation is above all. The nationalist approach shaped the Congress government and over a period of time, the Congress also exposed the workings of the Raj administration in a day-to-day basis during the governance. The Congress government had taken some decisions like the abolition of the zamindari system only in 1951 but had announced its abolishment much earlier during 1937–1939.[11] The zamindari system was an integral part of *Pax* India. "The expanded public discourse and the release of political prisoners significantly altered popular perceptions about the Raj's ability to criminalize protest and control the circulation of 'seditious' ideas. The Congress was successful, for the first time, in undermining the Raj's authority in *institutional* terms."[12] Throughout this shift, "the steel frame was put to considerable strain, and the authority of the district officer was seriously challenged.

[8]Ibid.: 152.

[9]Ibid.: 207.

[10]Ibid.: 207.

[11]Ibid.: 278.

[12]Ibid.: 279.

Perhaps the worst affected were the policemen who found their position considerably diminished. This loss of authority resulted in the policemen's under preparedness to perform their primary function-dealing with serious crime."[13] This increase in crime caused considerable anxiety to the government. Kudaisya points out, "The Congress ministers, of course, claimed that it was not crime but merely its reporting that had increased. G.B. Pant claimed that crime statistic had shot up because the public now did not have to bribe the police to file a complaint."[14]

The role of the Congress was pivotal in undermining the power of the *Raj*. "Once the Raj devolved the functions of provincial government to popularity elected ministries, the Congress successfully captured the state's institutions and undermined colonial authority, not by mobilization, but in a subtle manner-by catching the tiger by the tail!."[15]

In fact, by the early 1940s, the stage had been set for the "colonial heartland" to experience a dramatic political transformation. Political conditions were such that UP showed much resistance to the running of the colonial war effort.[16] UP had shaped the independence movement and Congress Party brought the British Raj at the negotiation table by giving them only one option—the freedom of India. The role of UP politics was evident in the making of independent India and gave enough administrative experience to the Congress Party about how to rule and address opposition political parties over an issue. The "colonial heartland" of course shaped the discourse administratively and politically as well.

The "Postcolonial heartland" discourse shaped UP in the contemporary period when politicians discussed the name of the state and contemplated naming it "Devagiri," "Himalaya Pradesh," "Aryavarta." But finally in the Constituent Assembly meeting, they came to a consensus and named the state "Uttar Pradesh."[17] In postcolonial UP, the politics of Muslim league either waned or the popular leader left for Pakistan and the rest of the population settled for the Congress Party in terms of their leadership. The issue of mother tongue was discussed amongst politicians variedly and, after long consultation, Hindi was adopted as the mother tongue of the state over Urdu.

The shape of UP has been moulded at various junctures to bring it to its present shape. In both colonial and postcolonial "heartland" discourse, the role of the Congress Party is evident for political reasons and the Congress Party and its politicians shaped the state to bring it to the present socio-politico stage.

---

[13]Ibid.: 279.

[14]Ibid.: 284.

[15]Ibid.: 286.

[16]Ibid.: 336.

[17]Ibid.: 351–9.

3.2 The Contemporary Identity of Heartland

The identity of the "heartland" has been shaped and reshaped over time and some issues emerged as bearing the symbolic identity of the "heartland" viz. caste politics (Dalits politics), communal politics (Babri Masjid–Ram Mandir issue) and legislative or electoral politics. The nature of caste politics was shaped in the form of "Dalits Politics" under the leadership of Kanshi Ram and Mayawati. The nature of Dalits politics was further reshaped after the demise of Kanshi Ram when Mayawati assumed the leadership of BSP and made space for "other" caste people to achieve her political ambition in order to leave a political imprint on national politics. Therefore, she astutely carved a political plan along with Satish Mishra to bring Dalits and Brahmin people onto a single political platform under the "social engineering" plan in 2007. However, the same plan could not be revived during the 2012 Assembly elections. The Dalits caste politics provided a new identity to the state with respect to both national and regional politics.

"The CSDS post poll survey findings show that the satisfaction of people with the BSP government was exactly the same as the satisfaction expressed by voters with the SP government in 2007 when a similar survey was conducted (Table 3.1)."[18]

In a survey CSDS finds that

> Mayawati-led BSP government it seems was also as unpopular as the previous SP government led by Mulayam Singh Yadav. When voters were asked whether the government should get another chance to rule, nearly half of them (48%) said no and only 27% were willing to give it another chance. When the same question was asked during a similar survey in 2007, 47% had wanted the then Mulayam Singh government to go and 37% had said it should get another chance (Table 3.2).[19]

However, the popularity of Mayawati in comparison to 2007 was diminished and the wide socioeconomic gaps in the state became the reason for defeat of BSP in the election, when the number of BSP MLAs could not reach the triple digit number.

The legislative strength of the state is 80 Lok Sabha Members of Parliament (MP), 31 Rajya Sabha MPs, 404 Ministers of Legislative Assembly (MLA); including one Anglo-Indian person and 100 Member of legislative council (MLC) which have made the state a politically important state in the country.[20]

---

[18]Kumar et al. (2012).

[19]Kumar et al. (2012).

[20]"The Legislative Assembly for the United Provinces was constituted for the first time on 1 April, 1937 in accordance with the Government of India Act, 1935. The strength of the Assembly as stipulated under the Act of 1935 was 228 and its term was five years. After Independence, the Legislative Assembly met for the first time on 3 November, 1947. Under the provisions of the new Constitution, the strength of the UP Assembly was fixed at 431 which was later revised to 426. After reorganization of the State on 9 November 2000, the strength of the Legislative Assembly has become 404 (When Uttrakhand become a separate state from UP) including one nominated member to represent the Anglo-Indian community. The term of the Legislative Assembly is five years, unless dissolved earlier" [http://legislativebodiesinindia.nic.in/States/UttarPradesh/upassembly-w.htm, accessed, March 16, 2013].

Table 3.1 Level of satisfaction with the Incumbent Government (2007 and 2012)

Satisfaction with Incumbent Government	SP Government 2007	BSP Government 2007	No in 2012 (in round off figure)
Satisfied with performance	49	49	3576
Dissatisfied with performance	35	39	2860
No opinion	16	12	855

Source CSDS, 2012 (EPW, April 7, 2012. Vol. XLVII NO 14)

Table 3.2 Popularity of Incumbent Governments in assembly elections (2007 and 2012)

Should the Government get another chance?	SP Government 2007	BSP Government 2012	No in 2012
Yes	37	27	1968
No	47	48	3514
No opinion	17	25	1808

Source CSDS, 2012 (EPW, April 7, 2012. Vol. XLVII NO 14)

Apart from legislative strength, there are some constituencies which have become very high-profile constituencies over time due to political reasons viz. Amethi, Rae Barelli, Maipuri and Lucknow. But during the recent General Elections of 2014, two constituencies' for examples Varanasi (Narendra Modi) and Azamgarh (Mulayam Singh Yadav) also became high-profile constituencies when both these leaders decided to contest elections from these respective constituencies to expand their political reach amongst people in the region. Giving a party ticket to these leaders from respective constituencies has somehow strengthened the conventional hypothesis that until a political party takes control of UP, it is not an easy task to assume power in Delhi. During the 1999 Parliamentary election, BJP got 29 seats and around 27.64% of the voting share. Therefore, BJP was able to form the NDA government under the leadership of Atal Bihari Vajpayee. The party also put forth a political plan to make Narendra Modi contest from Varanasi (Purvanchal) to make an impact on the voters in the region and expand horizontally by assuming control in the adjoining state of Bihar. During 2004 and 2009, in the Parliament elections, the Congress Party, SP and BSP, all did well as compared to BJP, allowing the UPA to remain in power for a decade. While in the 2014 Parliament elections, BJP won 71 seats out of 80 seats in the state (Appendix B.1). The electoral politics of course shaped the politics and made the state, the "heartland."

3.2.1 *The Making of Heartland Through Caste Politics*

In UP, both caste and communal politics are prevalent. The inherited caste factor has shaped the caste politics in the elections. The role of caste in electoral politics is evident as it reflects a caste configuration in the state and therefore caste is considered an important factor in the elections. The communal politics was inherent during the Babri Masjid–Ram Mandir controversy and with time it caused several communal clashes in the state and outside it in various capacities. Both 2Cs (Caste and Communal Politics) became the identity of the state over time.

The emergence of Dalits politics was shaped around the following Dalits ideologue. Dr. Bhimrao Ambedkar, Jyotiba Phule, Narayan Guru, E.V. Ramaswamy Periyar, Savitribai Phule and others laid down the ideological foundations of a Dalits-Bahujan movement in India. Charu Gupta examines the Dalits' history in northern India and writes

> The past three decades particularly have seen a flourishing of popular Dalits literature, pamphlets and books, which have emerged as a critical resource for deeper insights into Dalits politics and identity.[21]

While in UP, the Dalits politics shaped around Ambedkar (Baba Saheb), Kanshi Ram (Shaheb) and Mayawati (Behenji), who were treated like deities.[22] The installation of statues of these leaders in both Lucknow and Noida parks (Dr. B.R. Ambedkar Samjik Prerna Sthal and Rastrya Dalits Prerna Sthal), was an extension of that symbol when these leaders were treated as deities in society. Tiwari and Pandey argue that "for BSP's mass-base, the trio symbolized the ideology, vision, hope, leadership, agenda and strategy of the party." He further infers that BSP was Dalits's ideological party and the founder leader of BSP was Kanshi Ram, who professed the Dalits ideology in quite a "dictatorial authority in BSP."[23] However, Ghansyam Shah and Sudha Pai argue that

> It is in this changed context of a combination of factors that one has to locate the new agency among Dalits. The new class of political entrepreneurs that has emerged from amongst the ex-untouchable communities used the idea of "Dalits identity" and mobilized the SC communities as a united block on the promise of development with dignity. Some of them, such as Kanshi Ram and Mayawati, have been quite successful in doing so.[24]

The conundrum of caste politics can be understood through the discourse in the form of the slogans (Appendix B.5) that shaped the Dalits politics in the state. The politics has been played in the name of caste and in fact, caste has become the vehicle to fulfil the political aspirations of politicians. As Tiwari and Pandey point out, "Rhetoric is an important part of politics and nothing captures it more than the slogans of political parties."[25] The caste discourse is often expressed through various acts including slogans when supporters prefer to speak during public gatherings or address the caste struggle in society. During the 1980s and 1990s, Dalits politics can be understood through the slogans of the parties. For example,

---

[21]Gupta (2007).

[22]Tiwari and Pandey (2013, p. 163).

[23]Ibid.: 163.

[24]Jodhka (2010: 163).

[25]Tiwari and Pandey (2013, p. 163).

Tiwari and Pandey argued that the initial slogans of BSP were highly derogatory towards upper caste, likes "*Tilak, Taraju, Aur Talwar Inko Maro Jute Char*" (Brahmins, Baniyas and Thakurs should be beaten with shoes).[26] *Tilak* is a symbol of Brahmins (practicing priest by caste), as Brahmins use sandalwood/vermilion on their forehead. *Taraju* (weight balance) is considered the symbol of the Baniya community, whose main occupation is trade. *Talwar* (the sword) is considered the symbol of Rajputs, a warrior clan. The supporters of BSP often shouted this slogan in public gatherings without any fear. Instead they felt proud to do so. The slogan designates a particular period when the Dalits movement was on a warpath with other ideologies in society.[27]

The change of slogans represents a political shift to some extent, as can be seen when people started shouting, "*Hathi Nahi Ganesh Hai, Brahma Vishnu aur Mahesh Hai*" (the elephant represents Ganesh, and also the trinity, all Brahminical gods, who form the trinity of gods in Hindu mythology) and showing the coalition—the Brahmins and Dalits.[28] The BSP's party symbol, the Elephant, was represented in such a manner that it seemed like the Hindu Gods—Brahma, Vishnu and Mahesh had to pacify the relationships through symbolic interpretation.

Tiwari and Pandey also examine this slogan and say, "This slogan signified a shift in the politics of BSP as earlier it had always taken a hostile stance towards the Brahmins who they had termed as the upholders of the discriminatory caste system."[29] This was constructed to pay respect to Brahmins for political reason and was a part of politics as well. Interestingly, they argue, "Terrified of BSP's moves aimed at poaching its upper-caste vote base, the BJP resorted to a counter-attack. Playing on the caste ego of the Brahmins, it bounced back with "*Pandit Nahi Chamar Hai, haathi Par Swar Hai*" (the one who rides an elephant i.e. supports BSP is not a Brahmin, but a Chamar)—the slogan however failed to make much difference to the Brahmins who continued to flock towards the BSP."[30]

The discourse of Dalits politics changed over time and symbolized an era of politics where the Dalits leadership emerged as forerunner of UP politics.

---

[26]Ibid.: 189.

[27]Mayawati of Bahujan Samaj Party coined a vitriolic slogan *Tilak, Tarazu Aur Talwar Inko Maro Jute Char* (The three upper casts should be kicked) [http://www.languageinindia.com/oct2004/fatihiadvertising1.html, accessed February 24, 2013]. The BSP has come a long way since the days of its previous apolitical avatars the Bahujan and Minority Communities Employees Federation (BAMCEF) and the Dalits Shosit Samaj Sangharsh Samiti (DS4). Its slogans too have evolved along this sojourn. From "Tilak, taraju aur talawar, inko maro joote char" (It called upon the Dalits and backwards to shoe the brahmins, the banias and the thakurs) to "Hathi nahi Ganesh hain, Brahma, Vishnu, Mahesh hain." (It is not just an elephant, it is Lord Ganesha, It is Lord Brahma, Lord Vishnu and Lord Mahesh), the BSP has traveled quite a distance [http://timesofindia.indiatimes.com/india/Parties-come-out-with-catchy-slogans/articleshow/4391703.cms, accessed February 24, 2013], see, Tiwari and Pandey (2013, p. 189), see, http://shodganga.inflibnet.ac.in/bitstream/10603/4286/13/13_chapter%206.pdf, accessed February 21, 2012.

[28]Tiwari and Pandey (2013, p. 189).

[29]Ibid.: 189.

[30]Ibid.: 190.

Surendra S Jodhka discusses this, "The electoral victory of BSP in UP in 2007 also pointed to the role of the political leadership, and their ability to build viable cross-caste alliances." Jodhka further adds that "according to some commentators, it is the leaders and media experts who present/analyze electoral politics in caste terms, and make caste appear to be the single determining sociological variable in electoral politics. While the BSP is popularly seen as political party of the Dalits and is led by a Dalits woman, in its electoral mobilizations 'it did not pay too much attention to caste arithmetic and it did very well by imaginatively bringing a coalition of interest between different groups.'"[31]

The politics of "social engineering" of Mayawati was perhaps the biggest political event of 2007 in UP. The "social engineering" of BSP was in fact a political experiment when BSP successfully brought Brahmins and Dalits together for a political cause, which had never happened before in Indian politics. Mayawati executed the social engineering policy along with her trusted accomplice Satish Mishra, who became instrumental in the execution of the policy. Moreover, when Mayawati took over the charge of the party affairs and surprised the party members when she made the announcement for every caste and community to be included in the folds of BSP, she experimented first with Brahmins and opened the way for other castes. In the prevailing atmosphere, Mayawati took a political risk and played a political gambit of transforming the BSP from the party of Dalits to the "Sarvajan Samaj" or "Samta Muluk Samaj," which was one of her initial political understanding where she allowed some scope for people belonging to other castes. The "social engineering" policy of BSP paved the way for other forward caste people like Brahmins, Thakurs, Baniya and Rajputs. However, she judged the spatial political mood to harness a sense of political opportunity; therefore she went ahead along with her political accomplice Satish Mishra who helped her achieve the vote share in the name of "social engineering" (Table 3.3).

However, Vivek Prahladan finds that during 2012 UP election, "BSP given a higher number of seats to the Brahmin-Rajput combine than Dalits underscores the plausibility of the idea that such regional ethnic parties cannot intuitively rely upon their core constituencies."[32]

The slogan became instrumental in pacifying the relationships or the ways in which the political relationships are constructed. The political and ideological shifts can easily be accessed with the help of 2007 Assembly elections, which became instrumental to cement the relationship between the forward classes (Brahmins, Thakurs and Rajputs) and the Dalits. However, slogans became a medium to gauge when the political ideology was hostile and when the relationship was cordial and people could work together under "social engineering." The caste politics is evident in the state and often political parties draw lines to denote the population as a mere caste representation. Mayawati, herself, is a product of caste politics and she

---

[31]Jodhka (2010, p. 165).

[32]Prahladan (2012).

Table 3.3 Caste-wise seats contested and won by BSP in 2007 assembly election in UP

Castes	Contested	Won
Brahmins and other forward Castes	139	51
Dalits	93	62
Muslims	61	24
Thakurs	50	18
OBC's	110	51

Source http://shodhganga.inflibnet.ac.in/bitstream/10603/4286/13/13_chapter%206.pdf: p. 235, accessed February 21, 2012

has managed a Herculean task in shaping the "social engineering" and demonstrated her political skills across political parties.

In a survey conducted by CSDS, it was revealed that "both SP and BSP had to some extent lost its traditional voters. The vote share of the Yadav and Muslim community and the Jatavs were reduced to 6 and 24% respectively."[33] In a CSDS survey "among Brahmins for instance the SP secured 19% of the votes, up 9 percentage points since 2007. Among *Rajputs* its vote share went up by 6% and among *Kurmi* and *Koeris* its votes more than doubled. Not just that, SP gained across categories of age, education, gender, class and locality. Among men, women, poor, rich, rural and urban voters, SP's gains were in the range of 3–5 percentage points since 2007. While the BSP lost a significant chunk of its core, it seems to have held on to some of its gains among upper castes and Muslims in 2007. However among lower OBCs who had also given BSP an advantage in 2007, the party lost 11% votes. It's not just the BSP, but the BJP too which lost its grip over its traditional voters, namely, the Brahmins, Rajputs and other upper castes. While the party came down 6 percentage points among Brahmins, the losses among Rajputs and other upper castes were much bigger at 17 and 24 percent points, respectively."[34] BSP lost its traditional voters as compared to SP, while SP gained votes from forward and lower caste people from both rural and urban areas.

The role of caste in UP politics has been described in the caricature (Image 3.2) and it represents a "TOMB" where "T" stands for Thakurs, "O" for OBCs, "M" for Muslims and "B" for Brahmins. In UP politics, all these castes play an important role and in some constituencies, they even decide the candidate's future. Thakurs and Brahmins comprise a significant number in some constituencies and somehow decide the fate of the candidates. The Muslim population constitutes around 18.5% in the state and in some constituencies; it is more than 30% and is considered a deciding factor.[35] However, the OBCs vote bank politics is also considered pivotal

[33]Kumar et al. (2012).

[34]Kumar et al. (2012).

[35]See, Census 2011.

Image 3.2 Coronation of caste. *Source* TOI. 6.2.2012. p. 13 (Dance of Democracy)

during elections. Therefore, BJP promoted Uma Bharti to attract OBCs voters and women voters as well. Uma Bharti chose to file her nomination from Charkhari constituency (Bundelkhand) wherein more than 50,000 people of Lodh community live in the constituency and Uma Bharti belongs to the Lodh community as well.

The caste conundrum can be understood through the narrative of a villager who openly addressed the social apathy prevalent in the state and said that Putti Lal, a Dalits, lost his job as a cook in the primary school in Harak block in Zaidpur (SC) constituency in Barabanki district as the children from the upper castes refused to eat the food cooked by him.[36] Thus, caste is inherent in the social structure and society is expected to behave accordingly.

The subject of politics is never the same; rather it changes constantly and varies over space. The politics of UP is often known for caste and communal configuration but over time, both caste and community have been conceived as political discourse and have repeatedly shaped the spatiality of UP.

[36]For detail see under news headline 'Poll Plan: Caste shadows poll issues in UP' on 4.2.2012 on The Hindu, http://www.thehindu.com/news/national/other-states/poll-plan-caste-shadows-poll-issues-in-up/article2857849.ece, accessed February 4, 2012.

3.2.2 *The Making of Heartland Through Communal Politics*

The communal politics is not new to UP politics but the location of Babri Masjid and Ayodhya dispute and subsequent politicization of issue and widespread discussion, not just within the state but nationally, has created a general perception in the public sphere and has given UP a reputation for being a sort of breeding centre for communal clashes and communalisation of politics. This study is not intended to relive the past to discover the reasons for these clashes; rather it explores the reasons why the state is vulnerable to such frequent communal clashes and communalisation of state politics and polarization of the society.[37]

The communal clashes of UP have been looked at through various dimensions (Rajagopal 2001; Hasan 1996; Punyani 2003). But this issue has played an important rolc in polarizing thc socicty, particularly during election seen and commonly understood. These communal clashes are usually as being political acts, which are deliberately organized to win political edge over the opposition. The communal politics did not stop even in 2012, after the new political regime. *NDTV 24*7* telecast a news piece on September 18, 2012, about such communal clashes and concluded that in barely six months before Akhilesh Yadav was sworn in as the youngest Chief Minister of the state, six communal clashes took place in various parts of the state.

> In Bareilly district, some localities were under curfew for days. In Ghaziabad, the most recent, six people died in police firing after a mob turned violent hearing rumours about a copy of the Quran being desecrated. So far, 14 people have died in the violence across the state. Many more have been injured and property destroyed by angry mobs. Most of the riots have taken place in the last three and a half months. In June, violence broke out in Kosi Kalan near Mathura over a place of worship. Four people died and several vehicles were destroyed. In Partapgarh, houses of Muslims were burnt down after a Dalits girl was raped and murdered. Local villagers began rioting to protest police inaction. Bareilly district saw rioting twice in two months, both times over communal issues and curfew that extended for over a month. Moreover, the latest was in capital Lucknow and Allahabad, this time protests against the "atrocities" against Muslims in Assam and Myanmar.[38]

The recently held Muzaffarnagar communal clashes (July–August 2013) again raised questions about his ability to run the state, as more than 40 people died in the clashes. After these clashes, the political parties started mounting allegations at each other and the SP accused BJP for creating disturbances in the districts, while BJP hit back and placed the onus on SP and termed it as a complete failure of the government to check such incidents. While the political party an opposition was tagged as being insensitive and playing for political gain. However, the communal

---

[37]See news for details under headline 'Jaitley: Congress creating communal divide in UP' on 13.2.2012 on TOI, http://timesofindia.indiatimes.com/india/Jaitley-Congress-creating-communal-divide-in-UP/articleshow/11866302.cms?, accessed February 13, 2012.

[38]For more see, http://www.ndtv.com/article/india/six-riots-in-akhilesh-s-six-months-as-chief-minister-in-uttar-pradesh-268721, accessed, September 18, 2012.

Table 3.4 Number of communal riots and casualties between September 1 and November 20, 1992

S. No.	State/UTs	No. of riots	Casualties
1	Andhra Pradesh	4	2
2	Assam	1	7
3	Bihar	8	19
4	Delhi	1	8
5	Gujarat	26	99
6	Karnataka	22	88
7	Kerala	2	3
8	M.P	5	21
9	Maharashtra	3	4
10	Rajasthan	13	52
11	Tamil Nadu	1	6
12	U.P	28	224
13	W.B	2	6

Source Punyani (2003: 140)

clashes during 1992 did indicate the polarization of the society to a certain extent (Table 3.4).

Paul Brass argued that

> Hindu-Muslim communal riots have been an integral part of the political process in modern India since the 1920s. Although, since then, there has never been an extended period of time when Hindu-Muslim riots have not occurred somewhere in India, there have been times when they have occurred in waves or chains that have covered large parts of the country, in the post-Independence period notably during partition and before and after the great militant Hindu mobilization in the late 1980s that persisted until the destruction of the mosque at Ayodhya on December 6, 1992.[39]

The Hindu-Muslim clash in UP is not new as the genesis of this hostility lies in the Babri Masjid–Ram Mandir disputes where both communities claimed the site on grounds of religion. On several occasions, both communities clashed with each other and supported the issue for political gain. Meanwhile, the communal clashes of Muzaffarnagar exposed the contentious role of politicians who played a suspicious role in the episode.[40] Thus, political parties broke the cordial relationships between communities which had been intact in the districts where Jats and Muslims were living together for years.

Many communal clashes have taken place in the state over the years but it is not necessary that each communal clash has direct association with the Babri Masjid and Ram Mandir issue. Rather in most communal clashes, local issues have played

[39]Brass (2003, p. 366).

[40]See news for details under headline 'UP probes role of politicians in Muzaffarnagar riots' on 9.9.2013. HT, http://www.hindustantimes.com/india-news/lucknow/up-probes-role-of-politicians-in-muzaffarnagar-riots/article1-1119617.aspx, accessed September 9, 2013.

an instrumental role. The Hindu-Muslim unity consequently has widened over a period of time in the state.

3.2.3 Relations of Power: The Centre and the Heartland

Media often discusses the political strength of UP in terms of Centre–State power correlation. The proposition of media is based on the development of political pattern since the first Assembly elections that were held in 1952 in the State. The Congress Party emerged as being victorious and Govind Ballabh Pant assumed the charge by becoming the first Chief Minister of the state in independent India. Subsequently, in the central government, Congress Party assumed the office under the leadership of Jawaharlal Nehru as the first Prime Minister of independent India. At both the Centre and the state, the Congress Party was in power till April 3, 1967 when Charan Singh became the first non-Congress Chief Minister of the state and assumed the office for ten months and again for eight months from February 18, 1970 to October 1, 1970. This period was considered an exception when Congress Party was not in the power in the state but was in power in the central government. Post emergency, when elections were held, the Congress Party lost the elections and remained out of power for nearly three years. During that period, Ram Naresh Yadav and Banarsi Das of Janta Party ruled in the state. While at the Centre, Morarji Desai and Charan Singh of the Janta Party ruled during that period (Appendix B.4).

During the 1990s, the emergence of regional political parties was witnessed and coalition government was formed when Mulayam Singh Yadav assumed office in UP and at the Centre, the National Front came to power under the leadership of V.P. Singh with the support of BJP. But the government fell in both places—in the State and at the Centre. Thereafter, the United Front, NDA and UPA I & II formed different coalition governments backed by the support of political parties in power at the state level (Appendix B.2).

The media emphasizes and covers news related to particular constituencies that emerge as power centres in Indian politics as they are mostly associated with prospective Prime Ministerial candidates. Out of 13 Prime Ministers, the country has seen so far, 8 of them belong to UP and UP comprise 80 Parliamentary seats, which is around 15% of the total Lok Sabha seats. For example, Amethi, Rae Bareilly, Mainpuri, Rampur, Varanasi, have all become key constituencies in the present Lok Sabha elections and reiterate the predominance of Uttar Pradesh in National politics. It also demonstrates how media in particular has transformed these regional constituencies into major constituencies through its wide coverage (Appendix B.3).

3.2.4 Producing the Heartland Through Its Electoral Spaces

There are 403 Legislative Assembly seats and, 80 Lok Sabha seats or constituencies apart from 31 Rajya Sabha seats and 100 Legislative Council seats. But only a few of the Lok Sabha/Legislative Assembly seats are popular across the nation and even within the state. Why is this? How has this happened and who is responsible for it? The answer rests somewhere out of the state. How has an electoral space or constituency become pivotal over time and why? Everyone eyes this electoral space enthusiastically and it is even considered a base to draw various political conclusions (Table 3.5). For example, people are more interested in knowing about the political leadership in places such as Amethi, Rae Bareilly, Mainpuri, Azamgarh, Rampur and Varanasi even more than the leaders in their own constituencies. The reason is simple. These constituencies belong to powerful party leaders or they are considered as the face of the party in the elections or is there something else beyond these reasons that makes these electoral spaces unique amongst political spaces?

Painter (1995) argues for politics as social practice and these social practices usually comprise of two aspects—one is *material* and the second is *discursive*. The former is about the organization and the use of materials like absolute space as a base; while the latter is about ideas, language, symbols and meanings, to interpret the origin. The material cannot be separated from the discursive, but they are not the same thing. It is also possible that one can give material aspects more emphasis over discursive aspects, or vice versa. These constituencies were chosen for their voting familiarity and voting associations to the particular parties, and thereafter, this space was declared as a safe space to move (politically) without hindrance. For this study, material aspects are about the absolute space being used specifically as electoral space. The absolute space has been nurtured over time in such manner that a particular political party can take an edge over other political parties. The discursive aspects are about the leaders and their associations with these political

Table 3.5 Electoral spaces of UP

Prime Minister of India	Constituency	Prospective Prime Ministerial candidates	Constituency
Jawaharlal Nehru	Phulpur	Sonia Gandhi	Rae Braelli
Lal Bhadur Shastri	Allahabad	Rahul Gandhi	Amethi
Indira Gandhi	Rae Braelli	Mayawati	Akabarpur
Charan Singh	Bagpat	Mulayam Singh Yadav	Mainpuri/Azamgarh
Rajiv Gandhi	Amethi	Narendra Modi	Varanasi
V.P. Singh	Fatehpur		
Chandra Shekhar	Ballia		
Atal Bihari Vajpayee	Lucknow		

spaces particular. Of course, Amethi and Rae Bareilly became the bastion of Congress Party and Mainpuri became the bastion of SP over time. Atal Bihari Vajpayee was elected many times from Lucknow and the place was considered a favorite seat for BJP candidates even after his retirement from active politics. These constituencies have been developed very carefully to reflect the party's views, statements, symbol over time.

Joe Painter quotes Michel Foucault that "the meaning of language is not transparent and immediately obvious. Words, statements, symbols and metaphors and so on, mean different things in different contexts."[41] The meaning of this particular statement depends partly on who is speaking and how it is being said, but also how it fits in wider contexts. Over time, these constituencies have developed a high stature in the country and people are very concerned about these electoral spaces because of the political discourse that is supposed to be set from these spaces.

Take for an example the slogan *Garibi Hatao* (Remove Poverty), which is still intact as one of the familiar slogans of the Congress Party since the 1950s. The nomenclature has remained the same except some prefix and suffix were added or removed over time but the content remained the same. The slogan addresses the issue of poverty which is considered as one of the major social lacunae. The issue has been discussed on various platforms at length but during election time, the word becomes contentious across political parties, with all kinds of connotations. Over time, many people who were poor became rich, and to some extent vice versa. The "discourse" is how a word "Garibi" (poverty) becomes popular amongst people as well as political parties. Sometimes, the slogan has also been used to trigger a reaction against the Congress Party and has been interpreted as remove "poor" instead of "remove poverty". Nonetheless, the connotation is quite popular across the intelligentsia, politicians and social activists. The discourse of "Garibi Hatao" has been used in these constituencies by the Congress Party in order to fulfil the people's aspirations while interestingly; the opposition also used the same slogan against the Congress Party government.

Both Amethi and Rae Bareilly constituencies have been given extra care by the Congress Party and the association of the Nehru–Gandhi legacy has remained intact with these constituencies and has been transferred to the next generation of leadership with a sense of attachment and care. Over time, these constituencies have provided a sense of comfort to their leaders and thus, these constituencies have been given to the leaders of the political party so that they may be elected without any political obstacles. Thus, the constituencies have been nurtured under material and discursive aspects in order to reflect a safe haven for their leaders. Choosing districts/constituencies is indeed a hard task which has been done under material aspects and the space over time is groomed to reflect the direction which suits the concerned political party. Once the space is identified as a future political base, the discursive aspects are implemented to suit political decisions. However, the constituencies become popular across the section of people through various mediums,

[41]Painter (1995, p. 13).

including media, which made the constituency popular and political parties offered that constituency to their senior leaders for safe passage in the Parliament or in the Legislative Assembly as well.

Media brings everyday news from these constituencies to highlight the spatial perspective at myriad levels to make people aware of what is going in the concerned space. Media gives a lot of focus to these so-called pivotal constituencies and reports about this space through various programmes about what people are thinking about their leaders as compared to the opponents. Media being an institution circulates a political discourse and ignites the people's thought either by carrying out their institutional responsibility or to provide support to the concerned political party. The political discourse becomes contentious when media joins in the overall debate and discussion, whether it is "Garibi Hatao" (remove poverty) or some other political discussion with the concerned political party. Over time, political sensibility and political awareness amongst people has increased to a certain extent in order to know the overall view of politics. So, media has become one of the mediums to strengthen political practices and its affiliations across the spaces.

3.3 Representation of Heartland in Media Space

In the twentieth century, media is considered to be the greatest human invention and this discovery can equate this as one of the greatest epochs of human civilization so far. Media has become an inevitable phenomena and acts like a thinking machine that influences and shapes an individual to think along the line of thought that the medi proposes over an issue.[42] Media becomes a machine to leave an imprint about an issue over time. During the UP elections, the role of media was phenomenal for various reasons including the shaping and reshaping of the political discourse. Media raised the issue of caste and potential election issues including

[42]Media works in certain way to influence the potential election issues and therefore, sets an agenda and carves news in order to reflect a targeted direction to determine and influence audiences. During election days, the role of media is considered inevitable particularly when social media and TV media have become pivotal. "Agenda setting theory is the phenomenon of the mass media selecting certain issues and portraying them frequently and prominently, which leads people to perceive those issues as more important than others" (Wu and Coleman 2009: 776). Iyengar and Kinder infers Priming "changes in the standards that people use to make political evaluations" (quote in Scheufele and Tewksbury 2007: 11). "Priming occurs when news content suggests to news audiences that they ought to use specific issues as benchmarks for evaluating the performance of leaders and governments. It is often understood as an extension of agenda setting" (ibid.: 11). "Framing based on the assumption that how an issue is characterized in news reports can have an influence on how it is understood by audiences" (ibid.: 11). Media uses skilled techniques in order to influence audiences. Agenda-setting theory, priming and framing are considered tools which help the media, particularly news media, to influence their audiences. The news content, of course, admires geographical attributes and conceives media as a virtual space where geographical attributes are running at different representational form.

developmental issues which were discussed in prime time television programmes and daily editions of newspapers as well. However, media followed the state politics in various ways including mediated caste politics, the political landscapes of UP and the potential issues of UP Assembly election 2012 as well.

3.3.1 Mediated Caste Politics

Caste was a dominant political factor in the state while it has been used by everyone either by state or politicians or institutions in various ways. The word goondaism[43]/hooliganism has been often used for SP over the years in the media for creating disturbances. The mediated caste politics is an extension of such an understanding of how media makes such remarks which can be understood as mediated discourse. There are many regional and national political parties functioning in the state while only SP has been assigned the status of *goonda* party or *goondaism*[44] by the media through various programmes.

The political landscapes of UP became contentious in the media space when in 2003, SP took charge of the state and Mulayam Singh Yadav became the Chief Minister. This government survived till 2007, when Mayawati replaced him in 2012. The SP led UP was represented by the media as *goonda raj* or promoting *goondaism* and the word has been associated with the SP led UP government specifically.

How did the word *goondaism* become associated with SP and in what contexts were the meaning of *goondaism* portrayed in the media? The word *Goonda* means a hooligan who creates disturbances and unrest. The word is stereotypical and carries negative connotations. Interestingly, the word *goondaism* again resurfaced after the 2012 Assembly elections when the Samajawadi Party had won the elections in UP and talks of *goondaraj* days are making news across the media. Why the term *goondaraj* has been associated with the SP alone has never been debated on any platform. Why has this term not been used for BJP or BSP or any other political party so far, even though their government also failed to check criminal activities in the state? Why the Congress party was never tagged with this term when it ruled the state for so many years? Moreover, what does the term really mean? For the answers to these questions, we need to look at what the term *Goonda*/*Goondaraj* means in UP's social structure.

The meaning of *goonda* is not equivalent to a criminal. Criminals can turn *goonda*s and sometimes *goonda*s can become criminals. Criminals take to crime as a profession, as a means of livelihood. *Goonda*s do not practice *goondaism* as a

[43]See news for details under headline 'यूपी में 'उगाहीराज' और 'गुंडाराज' का बोलबाला' on 3.2.2012 on Rastriya Sahara.

[44]However for BSP opposition political party using words 'उगाहीराज' (collection) while for SP it is using 'गुंडाराज' (Goonda Raj or Goondaism). Hence both the political connotation was used across the media in order to signify both the political party in various capacity.

profession. They have other means of livelihood. Criminals prefer to life in hiding, while *goonda*s roam the streets openly with no fear of being caught. A criminal would not engage in activities that do not result in any monetary benefit for them. *Goonda*s may appear to be criminals but it is more about social authority than economic profit. Through few appropriate examples it is easier to understand the nature and characteristics of *goondaism*

> A Dalits/Muslim or anyone would park his bike at the roadside of a tea shop. A criminal may steal that bike or snatch it from the owner. The goonda would neither steal the bike and nor snatch. He would take the air out of the tyres and wait for the owner to come back. While the biker would look for a way out of this predicament, the goonda seated nearby, would laugh. They have thus earned a higher reputation for themselves.
>
> A young lady is walking down the street. A criminal would have his eyes on the purse or jewellery the lady is wearing, the goonda would have his eyes on the body of the woman. This would be followed by whistling which will have the woman run for cover. The goonda has made his point.
>
> A paan (chewing betel) may cost a couple of rupees in the street. A goonda would never pay for the same. If the paanwallah would ask for the money, he may be abused and slapped. From the next day, the paanwallah will not suffer any humiliation. The moment the goonda would arrive with his esteemed colleagues, the paanwallah would quickly hand him two paans.
>
> The tea stall owner, too, would prefer to serve the goondas free snacks and tea. As we can see, all these people in no way support the livelihood of the goonda. The goonda now has a new territory where citizens fear him. A criminal would do neither of the above. He would wait for the Sun to set and rob the paanwallah. And how does it become goondaraj? The answer is simple. Goondaraj is the rule of goondas.[45]

The physical territories of goondas are spread across the country but no one is bothered about it while media compares the incidents with the SP. The mediated politics can be understood with the help of a caricature produced by *The Times of India* (Delhi edition) the criminal activities flourishing in the state is associated with specific political parties (Image 3.3).

The caricature sharpens an understanding of current thinking amongst people. The cartoon is clear in its meaning and essence and represents the party symbol with something which can cause hooliganism and *goondaism*. The party symbol was restructured where all frames are fixed with barrels. The gun is a symbol of crime and can cause social disturbance. The caricature also reflects an understanding of media about the SP in the state and projects the fear of crime associated with the party. However, the whole story is led by three leaders talking amongst themselves and looking at the party symbol, a bicycle that has been redesigned in such a way that the party can manage the spatial promises that they made during election campaigns in their party manifesto in the cartoon. All three leaders are standing near a Welfare Project Monitor office where they are supposed to decide and think about starting projects (distribution programme), in order to fulfil the promises that they made in their speeches during election time. The party symbol

[45]For detail see. 'Urgent for change in daily'. The pioneer, New Delhi edition. Dated 24.3.2012.

Image 3.3 Representation of election symbol. *Source* TOI. 25.4.2012. p. 9

has been fitted with eight barrels and can fire from all directions, whenever required to do so.

The cartoonist Ajit Ninan employed by *The Times of India* (Delhi edition), represents the spatial understanding in such a contentious manner where party leaders are supposed to "implement schemes *bang* on time," and one of the men is carrying a gun. The way in which the gun has been associated with the SP and their leaders nudges us to draw the inference that the Party is promoting *goondaism* rather than checking it. In fact, media being an institution is free to make such associations as part of their independent reporting and mapping a regional party. But such remarks raise questions about the neutral and rational approach of media and thus, it seems as mediated politics.

3.3.2 The Political Landscapes of Uttar Pradesh

The political landscape of UP is heterogenous and every political party is looking for their political share. The political debates during 2012 Assembly elections was

Image 3.4 People's agony. *Source* TOI. 15.2.2012. p. 13 (Dance of Democracy)

discussed in order to address national and regional politics. "The debate on electoral politics has been restricted to the binary of identity or ethinic politics on the one hand and the other-the former as the politics of belonging (identity) and latter as that of belongings (roti, kapdaa, makan). These two generes of political practice are getting conflated within the contest between the regional (SP and BSP) and 'national' parties (Congress Party and BJP)."[46]

The political landscape of UP has been driven from national to regional politics over time and in between, four political parties have ruled the state. Amongst them, the Congress Party has ruled for the maximum period and other political parties have ruled either through coalition or for short periods—BSP ruled in 2007 and SP has been ruling the state since 2012 independently. The presence of BJP in state politics is noted during 1997–2002 when BJP formed the government in the state. These four political parties have ruled over the state and therefore, there is a feeling that they are supposed to work for the development of the state. The (Image 3.4) represents the "four cornered fight," where voters are scolding these political parties for not developing the state so far. In this caricature, the two men discuss that over six decades, these political parties have made a fool out of people and that the voters should "corner" them. The political landscape of the state has not been the same after 1989 or post Babri Masjid–Ram Mandir movement when BJP polarized the society. However, the political landscape of UP can be understood through the political phases in the state and the political shifts that took place when regional parties were given preference over national parties after the 1990s.

---

[46]Prahladan (2012).

Image 3.5 Rhetoric of Indian politics. *Source* TOI. 11.2.2012. p. 13 (Dance of Democracy)

During the 2012 elections every exit poll had predicted about the win of SP more or less meanwhile, SP emerged as a larger political party in the state. Across the political parties, it was speculated that the 2012 election is a multi-cornered contest and none of the political parties was in a position to claim the future government but informally, every party was staking a claim in forming the government. The political allegations and counter allegations became common political activities during the election campaign and every political party had its own understanding about this. The political landscape was turning into a new political discourse every day, a fact that was portrayed in media as well (Image 3.5). When political parties alleged that Rahul Gandhi never visits the state and rules through remote control; the politics around this political discourse provided political opportunities to every political party to comment something on the issue. Therefore, in an interview to the media, Rahul Gandhi conspicuously said that in actuality that none of the political parties have any such control over the state. He explained it and said, "UP is remote from any control, mine or anyone else's." The dynamics of politics in UP is broad wherein caste politics is considered an important element.

UP politics has seen many ups and downs since independence in India. The political landscape of UP is quite well-known across political parties and to the politicians as well. It shall be easier to decode the politics of UP if we shed some light on UP politics through three different phases in terms of the political rule—the Congress Party regime; Coalition politics regime; and Regional Party regime.

The British carved the North-Western Province in order to provide better administration and moved the administration from Allahabad to Lucknow. Post independence, the Congress party lead the region in every possible manner and thereafter ruled for almost four decades in UP without much political opposition. During that period, the Congress ruled over the state very smoothly except during the 1970s when the Emergency was imposed in the state despite strong opposition. The rule of the Congress party for such a long period and its misgoveranace provided an opportunity to some politicians to dissociate themselves from the Party and they separated to provide more political options to the people of the state. Although the emergence of regional parties gained momentum after the Emergency, it was the Babri Masjid–Ayodhya Mandir movement that provided ammunition to the regional parties to carve out their political space in the state. The role of *mandal* politics also played an important role in shaping the regional politics in the state.

Zoya Hasan charts this change and says, "The 1980s were a crucial phase in the transformation of UP politics. Four factors dominated the process of change: the alienation of rural producers, discontent of marginal groups in rural society, growing assertion of the backward castes, and the challenge of communalism."[47] During the 1990s, the nation was under crisis and was facing two major problems-economic instability with low economic growth and communal clashes in several parts of the country. Both these events changed the facet of Indian politics and as a result, people were pushed to think of alternative political options. This was the phase when Congress was almost replaced by regional political parties in the state. The political consequences were seen as the genesis of regional political parties, which subsequently replaced the Congress Party in the state. Both SP and BSP are to some extent successful in keeping the Congress Party off from the state over the years. The failure of the Congress party in the 2012 elections consider an extension of regional politics wherein people did not buy into the politics of a national party but rather preferred the regional party for various reasons.

After the Janta Party government in the 1970s, once again a coalition government was formed when Mulayam Singh Yadav[48] assumed charge in 1989 with the support of BJP. However, this government could not complete its tenure due to political tensions between the Janta Dal and BJP. Over time, there have been several coalition governments formed but none of them can be called successful as political tussles between political parties somehow derailed the coalition in the state. The coalition between BJP and BSP also ended on a disgusted note with both parties restraining from getting into any such pact in the future. The coalition between SP and BSP also ended with disruptive views and vows to avoid each other even outside the State. Although, the coalitions were geared towards keeping the Congress Party out of power, in which both the regional parties (BSP and SP) were partly successful; the fate of the Congress is still hanging and they were looking for a political miracle. However, the 2012 elections again failed to reinstate

[47]Hasan (1996, p. 84).

[48]Mulayam Singh Yadav found his political genesis during the JP movement and since then he became an instrumental regional leader in the politics of the UP. He became the CM in the year 1989 with the support of BJP but the government fell due to the reiterated demand of BJP to construct Ram Temple at Ayodhya, which caused the Janta Dal-BJP coalition in the state and in the centre also.

the Congress party, despite the elections being fought under the leadership of Rahul Gandhi.

Sudha Pai, in her examination of the political scenario in the state writes, "The BSP political regime starts when BSP contested elections for the first time in UP in the 1984 Parliamentary elections and the following 1985 Assembly elections. Between 1984 and 1989, the BSP could win no seats, but its base among Dalitss, especially government employees, grew."[49]

Over time, the vote percent of BSP increased incredibly. However by 1993, Mulayam Singh Yadav formed SP by breaking away from SJP (Samajwadi Janta Party) and by the 1993 elections, "the SP and BSP as two lower caste parties well positioned to jointly oppose the communal mobilisation of the BJP."[50]

However, Mayawati[51] was successful in assuming the post of CM of UP thrice through coalition, either with the support of SP and BJP respectively and once she formed a full-fledged government (2007–2012) in the state. The political wheel again turned when the BSP supremo diverted her political policy from "Bhujan Samaj" to "Sarva Samaj" and opened the door for people belonging to every caste to join BSP. Thus, she played the politics of inclusion in the form of "social engineering" where "social" and "caste" were experimented upon in a new political format by Mayawati in 2007. Across the political parties, there was a common understanding that in the present situation in UP, none of the regional parties can rule over the state singularly. During the 2007 Assembly elections, Mayawati secured 206 Assembly seats out of 403 total Assembly seats only after successful implementation of social engineering wherein Brahmin and Dalits voted together for BSP. However, the experiment of "social engineering" did not work for the second term but was rather criticized on ethical and moral sanctity of such political bondness and social alliance that it was considered political opportunism. Oliver Heath and Sanjay Kumar point out that

> In 2007 the BSP secured an overall majority, winning 206 seats and 30% of the vote. During the 2012 Assembly elections the BSP lost 4.5 percentage points in votes and lost 126 assembly seats. Though the decline of 4.5 percentage point votes may not look like a big loss; in a four cornered contest even a modest change in votes can have a big impact on seats. Moreover, of potentially great concern to the BSP, the party's popularity declined most amongst its core supporters. The BSP fared particularly badly in places where it had traditionally done well, such as in the assembly constituencies reserved for Dalitss. In the 2007 Assembly elections, the BSP won 61 of the 89 seats reserved for Dalitss, which corresponds to 68% of the reserved seats. This time, in 2012, the BSP won only 15 of the 85 seats reserved for Dalitss, which corresponds to just 18% (following delimitation the number of seats reserved for Dalitss declined from 89 to 85).[52]

[49]Pai (2002, p. 155–6).

[50]Ibid.: 162.

[51]She assumed the post of CM of UP four times. She reformed the political ideology of BSP and made space for "others" caste in order to promote *Sarva Dharma Sambhav* as well. She was popular leader among Dalits population and ideologue too.

[52]Heath and Kumar (2012, p. 41).

However, BSP lost the election badly and SP won the elections in 2012 with a huge majority and also received the overwhelming support of the Dalits.

National politics to some extent is manoeuvred by those sections of politicians who control UP politics. In post-2012 Assembly elections, both SP and BSP extended their outside support to the Central Government and on many occasions the SP bailed out the Central Government from political crisis. For instance, during the nuclear crisis, when Congress-led UPA-1 was supposed to fall, it survived due to the support of its allies. However, Congress-Party-led UPA-II government is successfully running after the support of both SP and BSP, while simultaneously maintaining political distance as well. Both SP and BSP were defeated by a huge margin by BJP in the general elections of 2014. BJP secured 71 seats out of 80 Parliamentary seats while SP got five seats of the family members including Mulayam Sigh Yadav (Mainpuri and Azamgarh) seat while BSP was unable to secure any seat and Apna Dal got two seats. Thus, once again the heartland proved that those who control the state of UP shall control national politics as well.

3.3.3 Locating Uttar Pradesh Assembly Election 2012 in Media Spaces

In a democracy, elections are considered as being crucial for political parties and people to elect their representative through electoral franchise. However, the SP won the election and Akhilesh Yadav became the youngest Chief Minister of the state in 2012. The Uttar Pradesh election was considered pivotal because along with Uttar Pradesh, four more states had elections held simultaneously but none achieved the status or attracted the media attention which Uttar Pradesh did.

Every election witnesses a set of election issues and draws politicians across political parties to address the issues earnestly. During the UP Assembly elections, people found some issue which was pertinent and reasonable in the state.

In a CSDS "survey (Table 3.6) revealed that price rise was the most important issue for voters (36%) during the election. Condition of roads emerged as the second most important issue (26%), followed by farmers' problems (14%). Corruption was an issue for 7% of the voters."[53]

But apart from the above-mentioned issues, media also set certain election "agenda setting"[54] and identified eight issues (images) which were considered potential issues to represent the state where media, the state and politicians/political parties followed these issues at best and these issues become contentious across political parties during the 2012 election campaign. One of the oldest national Hindi daily news channels, Zee News, had outlined eight potential issues and these issues served as a base to understand the mediated political discourse about UP Assembly

[53]Kumar et al. (2012).

[54]McCombs and Shaw (1987).

Table 3.6 Most important election issue in UP (in percent)

Price rise	36
Conditions of road	26
Farmer's problem	14
Unemployment	7
Corruption	7
Electricity, water, etc.	3
Development of state	2
Law and order	1

Source CSDS, 2012 (EPW, April 7, 2012. Vol. XLVII NO 14)

elections 2012. These eight issues were included from all possible sections of different political shades and were considered major ones even amongst political parties including the voters of the state. The mediated issues are[55]

1. Statues, Memorials, Parks and Elephants
2. Reservation for Minorities
3. Division of UP
4. Land acquisition stir on *Bhatta Parasual* agitation
5. NRHM scam and Khuswaha fiasco
6. Rahul Gandhi-A tireless Campaigner
7. The anti-incumbency wave against BSP
8. UP: The nerve centre of politics.

These issues to some extent covered almost all significant issues and were discussed by the state and media as well. Interestingly, since the Anna Hazare[56] agitation against corruption and implementation of *Jan Lok Pal*,[57] it became one of the contentious issues which media discussed particularly corruption and the nexus between criminals and politicians. However, corruption was one of the issues in the state election but was not considered as a major issue in the election but across media, the issue became more debatable and to some extent, the news was linked with UP based politicians as well. During the election coverage media often raised the issue of corruption and the nexus between criminals and politicians in various ways. However, media prepared detailed lists regarding the number of criminals, millionaires, women, graduates, youth who had managed to enter UP Assembly.

[55]For details, http://zeenews.india.com/state-elections-2012/up/issues.html/2, accessed October 28, 2011.

[56]Anna Hazare was a noted social worker who played an instrumental role in pushing the Indian government to pass the Lokpal bill. His Lokpal movement was discussed extensively in the media during his five day hunger strike at Jantar Mantar in April 5–11, 2011 onwards. However, he shaped the discourse on corruption and Lokpal issue in India.

[57]An ideological lokpal bill prepared under the leadership of Anna Hazare and his team members including Arvind Kejriwal and others. However, the government did not accept it fully but only followed some points from it.

Table 3.7 Media coverage of news

Source of news	Number of news		Total
	Hindi	English	
Newspapers	204	179	383 (51%)
TV news channels	256	120	376 (49%)
Total	460 (60%)	299 (40%)	759 (100%)

Source Primary Research Survey 2011–2012

During the UP Assembly elections, media produced and reproduced a number of images[58] to represent the state. These images were portrayed for political endorsement as well.[59] This study is aimed at examining the ways in which media portrays the space/place in their media space (newspaper and TV screen) through the news. This study collected around 759 news items regarding the UP Assembly elections of 2012 over a period of more than five months from both TV news channels and newspapers. Around 60% of the total news items were collected in Hindi language while 40% of them were in the English language. Around 51% of the news items were collected from which 204 were from Hindi language and 179 from English Language newspapers while 49 % % news items were collected from TV news channels in which 256 belong to the Hindi language and 120 to the English language (Table 3.7).

News regarding the UP Assembly elections 2012 appeared across media spaces in different ways. The news item "UP: The nerve center of politics," comprised the maximum number of news and was followed by other news items—"Rahul Gandhi: A tireless Campaigner; Reservation for Minorities; NRHM Scam/Kushwaha Fiasco; Statues, Memorials, Parks and, Elephants; Division of UP; Anti-incumbency wave against Mayawati and Land Acquisition stir in Bhatta Prasual." During UP election

[58]Every day we encounter political news. Bennett and Graber infer political news as: "what newsmakers (politicians and other political actors) promote as timely, important, or interesting; from which news organizations select, narrate, and package for transmission (via communication technologies); to people who consume it at a given time in history". News is something that is constantly changing the interaction of the trio (journalist, politicians and peoples). News is about sketching an image of particular place/space based on true incident/accident in order to subscribe who, when and whom the subject is concerned. Therefore, news considers place/space images to make the readers/audiences aware. "Place images have been systematically studied by geographers since the 1970s" (Adams 2009: 140). Being geographer, I am keen to know that what happened to the place/space with the help of media who brought us information in form of news, which I am considering in my study to explore the space through the lens of media. Prior to the assembly elections in Uttar Pradesh, which were held in 2012, the media disseminated information in the form of news, which can be classified into many categories to draw some inference.

[59]Technological determinism is a state of conditions when an individual act is somehow governed by technology and he/she completely depends on technology for various purposes. For example, a mobile phone becomes a mediated gadget when people feel comfortable for a purpose or for different reasons and utilize the available means (news etc.) in an optimum way. How did mobile phones become an instrumental factor to ensure the victory of BSP in the assembly elections of 2007? For further on this see Jeffrey and Doron (2012).

Table 3.8 Number of news appeared in Hindi and English language

S. No.	Issues	No. of news		Total (Hindi + Eng.)	Percentage of news of total news
		Hindi	English		
1	UP: Nerve center of Politics	231	124	355	46.7
2	Rahul Gandhi: A tireless Campaigner	96	81	177	23.4
3	Reservation for Minorities	29	37	66	8.7
4	NRHM scam/Kushwaha fiasco	34	13	47	6.2
5	Statues, Memorials, Parks and Elephants	20	22	42	5.5
6	Division of UP	15	16	31	4.1
7	Anti-incumbency against Govt.	27	01	28	3.7
8	Land acquisition stir on Bhatta Prasual	08	05	13	1.7
Total		460 (60%)	299 (40%)	759	100

Source Primary Research Survey 2011–2012

campaign the two news issues "UP: The nerve centre of politics" and "Rahul Gandhi: A tireless Campaigner" was quite popular and discussed in the media and almost comprise 70% of total media discussion. While rest of six issue discussed across the media was least in comparison to "UP: The nerve centre of politics" and "Rahul Gandhi: A tireless Campaigner" issue and comprise almost 30% of total news content discussed during UP Assembly election. The other three popular story discussed during the election was "Reservation for Minorities; "NRHM Scam/Kushwaha Fiasco" and; "Statues, Memorials, Parks and Elephants" and each story comprise more than 5% of total news content which appeared in both TV news channels and newspapers. While on following issues "Reservation for Minorities," "Statues, Memorials, Parks and, Elephants" and, "Division of UP," number of news appearing in English language were more. While the rest of the other issues appeared more in Hindi language media (Table 3.8).

Each media space has its own trajectory and covered the issues accordingly. *The Hindu* covered the maximum news regarding the issue of, "UP: Nerve Center of Politics" amongst the English language newspapers.

The Hindu is a south Indian centric newspaper and the newspaper is available only in a few metro cities in north India. The fact that *The Hindu* recorded the maximum news on the issue of "UP: Nerve Centre of Politics" establishes the fact that UP is a politically strategic state in the country. While *Rastrya Sahara* being a

Table 3.9 Number of news appeared in selected media spaces (in percent)

S. No.	Issues	Zee News	Aaj Tak	Rastriya Sahara	DD	NDTV 24*7	CNN IBN	The Times of India	The Hindu
1	UP: Nerve center of Politics	22 (25%)	55 (34.3%)	153 (75%)	1 (12.5%)	35 (55.5%)	19 (33.3%)	48 (38%)	22 (42.3%)
2	Rahul Gandhi: A tireless Campaigner	35 (39.7%)	30 (18.8%)	30 (14.7%)	1 (12.5%)	11 (17.4%)	19 (33.3%)	39 (31%)	12 (23%)
3	Reservation for Minorities	07 (7.9%)	16 (10%)	05 (2.4%)	1 (12.5%)	05 (8%)	03 (5.2%)	19 (15%)	09 (17.3%)
4	NRHM scam/Kushwaha fiasco	12 (13.6%)	15 (9.3%)	06 (3%)	1 (12.5%)	02 (3.17%)	02 (3.5%)	06 (4.7%)	03 (5.7%)
5	Statues, Memorials, Parks and Elephants	05 (5.6%)	08 (5%)	06 (3%)	1 (12.5%)	07 (11.1%)	06 (11%)	06 (4.7%)	03 (5.7%)
6	Division of UP	04 (4.5%)	08 (5%)	02 (0.9%)	1 (12.5%)	01 (1.6%)	06 (11%)	06 (4.7%)	03 (5.7%)
7	Anti-incumbency against Govt.	–	25 (15.6%)	01 (0.5%)	1 (12.5%)	01 (1.6%)	–	–	–
8	Land acquisition stir on Bhatta Prasual	03 (3.4%)	03 (1.9%)	01 (0.5%)	1 (12.5%)	01 (1.6%)	02 (3.5%)	02 (1.5%)	–
Total		88	160	204	8	63	57	126	52

Note News has been recorded from October 1, 2011 to March 7, 2012
Source Primary Research Survey 2011–2012

local newspaper of UP covered the maximum news pertaining to the issue of "UP: Nerve Centre of Politics." Every selected media space had covered the issue more than 30% except *Aaj Tak* which had covered the issue around 25% of the total news pertaining to the issue. News on "Rahul Gandhi: A tireless Campaigner," had appeared the maximum number of times in *The Times of India* amongst all English language newspapers while Hindi TV news channel, *Zee News,* covered it to the maximum (Table 3.9).

During elections, media covered all concerned issues and provided a platform to discuss the issue. Media was supposed to present political discourse in a way to provide a complete picture of the spatiality in political contexts. Such political discourse[60] is about the presentation of texts and contexts of politicians in the form of news.

3.4 Conclusion

In both colonial and postcolonial UP, the role of the Congress Party was instrumental in shaping the political discourse of the state—whether it was the issue of naming the state or deciding the mother tongue of the state. The role of the Congress Party was incredible particularly in giving physical and political shape to the state and until the last phase of 1990; the Congress Party was one of the biggest political players in the state. The sentiments of nationalism amongst people were driven by the Congress Party and in fact the Party led such sentiments against the

[60]Discourse simply means speech or communication or to engage in communication with others through a variety of means. The discourse analysis is an analysis of the way in which people communicate. The political discourse is simply about political discussion, when political parties discuss about political issues or other matters pertaining to the political understanding. My concern about the political discourse is the way in which media represents the space through news headlines where political texts and contexts are shaped and reshaped in quite "real" and "imagined" ways to bring a spatio-politico meaning. The political discourse is about the discourse of political leaders of different political parties who say different discourses on different occasions in their political capacity to adjust political understanding. "The overall idea of discourse theory is that social phenomena are never finished or total. Meaning can never be ultimately fixed and this opens up the way for constant social struggles about definitions of society and identity, with resulting social effects. The discourse analyst's task is to plot the course of these struggles to fix meaning at all levels of the social". Here, I am considering news as a political discourse which is neither open nor closed, but is rather shaping and reshaping the spatial discourse. A political understanding is posited through composite words where meanings are expressed in multilayer forms. It is like a 'capillary' notion of power, the concept of 'discursive formation' of Michel Foucault. Over period, subsequent authors have simplified the meaning of discourse (see Painter 1995: 13–14). The political discourse is here to substantiate the political knowledge when media is supposed to examine the same with a title or headline, is understood as political discourse. See Chap. 2 in this study to see how political discourse is established over a period of time and how the '*Garibi Hatao*' programme was established in our socio-politico space. Here, political discourse is about the news which started in some contexts where issue based subjects were discussed in myriad capacities. See Bennett (2007).

Raj. During British Raj, the Congress played an instrumental role to shaped the administrative institutions including police, judiciary and legislative process.

The political clock changed after the 1990s, when regional parties, primarily SP and BSP replaced the Congress Party and subsequently, the emergence of Dalits politics was evident in the state that shaped the country's politics during the 1990s and 2000s. Dalits politics took another turn when the BSP leadership allowed the inclusion of people belonging to other castes in their folds to expand the horizon of the party in the country. The political experiment was a task for politicians to move accordingly. The political strength was often shown in a way of centre–state correlation where the same political party ruled in the state as well as at the centre. Prior to the 1990s, the Congress was in the state and subsequently was also in power at the centre. While after the 1990s, when regional parties emerged, they started playing an instrumental role in the coalition government either under BJP or Congress Party leadership. Meanwhile, media portrayed the landscape of the state in various political shades particularly under the BSP regime from 2007 to 2012 in a negative way where frustration, anguish, corruption, law and order situation remained in the headlines during the election period.

Over time, UP emerged as a politically strategic state and political parties struggled to control it in the best possible way. The decision taken by Mulayam Singh Yadav and Narendra Modi to contest election in 2014 from Azamgarh and Varanasi respectively was indeed a part of that strategy to control maximum voters and expand the horizon of the party in the region. The decision of Narendra Modi to contest from Varanasi was considered a strategic political move to control UP politics to some extent because unless they had the control of UP, it would not have been an easy task for BJP to move forward in the national politics.

Media uses images to represent the election issues in the best possible way and over time these images becomes a symbolic "representation"[61] and acquires an identity to represent the politics in the state in mediated ways. The representation of the state as "Heartland" in the media through news will be discussed in the forthcoming chapters.

The next chapter discussed the three issues (images) out of eight issues wherein media was supposed to highlight the political landscape of the state in shape of

---

[61]Representation is about a symbolic identity either political or some other way. See, Lefebvre (1991). Lefebvre argues that "Representational space is alive: it speaks, it has an affective kernel or centre: Ego, bedW bedroom, dwelling, house; or: square, church, graveyard. It embraces the loci of passion, of action and of lived situations, and thus immediately implies time. Consequently it may be qualified in various ways: it may be directional, situational or relational, because it is essentially qualitative, fluid and dynamic" (1991: 42). The issues raised and discussed extensively during the election period in UP had to some extent shoved spatiality into something directional, situational and relational with spatial expectations. Of course, these issues are set to exemplify the spatial manifestations. "Representational spaces are symbolic works. These are often unique; sometimes they set in train 'aesthetic' trends and, after a time, having provoked a series of manifestations and incursions into the imaginary, run out of stream" (1991: 42). "Representational spaces, which are limited to works, images and memories whose content, whether sensory, sensual or sexual, is so far displaced that it barely achieves symbolic force" (1991: 50).

news across different mediascapes. Mediascapes discussed UP Assembly election-2012 as: news pertaining to the announcement of election schedule, election manifesto as political gimmick, use different methods to pull the crowd at election rally, political furor to bifurcate the state into four new states and, the installation of Statues, Elephant, Memorials and Parks in the state.

References

Adams PC (2009) Geographies of media and communication: a critical introduction. Wiley-Blackwell, Chichester

Bennett WL (2007) News: the politics of illusion. Pearson and Longman, New York

Brass P (2003) The production of Hindi-Muslim violence in contemporary India. University of Washington Press, Seattle

Gupta C (2007) Dalit 'Viranganas' and Reinvention of 1857. EPW. May 12, 2007:1739–1745

Hasan Z (1996) Communal mobilization and changing majority. In: Ludden D (ed) Making India Hindu: religion, community and the politics of democracy in India. Oxford University Press, New Delhi

Heath O, Kumar S (2012) Why did Dalits desert the Bahujan Samaj Party in Uttar Pradesh? EPW. July 14, xlvii(28)

Jeffrey R, Doron A (2012) Mobile-izing: democracy, organization and india's first "mass mobile phone" election. J Asian Stud 71(1):63–80 (February)

Jodhka SS (2010) Caste and politics. In: Jayal NG, Mehta PB (eds) The Oxford companion to politics. Oxford University of Delhi, New Delhi

Kudaisya G (2006) Region, nation, "heartland": Uttar Pradesh in India's body politics. Sage Publications, New Delhi

Kumar S et al (2012) Sixteen assembly elections in Uttar Pradesh. EPW. April 7, XLVII (14):80–86

Lefebvre H (1991) The production of spaces (trans: Nicholson-Smith D). Blackwell, Oxford

McCombs M, Shaw D (1987) Chapel Hill study. In: Iyenger S, Kinder D (eds) News that matters: agenda setting and priming in a television age. University of Chicago Press, Chicago

Pai S (2002) Dalit assertion and the unfinished democratic revolution: the BSP in UP. Sage Publications, New Delhi

Painter J (1995) Politics, geography and political geography: a critical perspective. Arnold, London

Prahladan V (2012) UP assembly elections: politics of 'belonging' or 'belongings'?. EPW XLVII (6):10–11 (February 11)

Puniyani R (2003) Communal politics: facts versus myth. Sage Publications, New Delhi

Rajagopal A (2001) Politics after television: Hindu nationalism and the reshaping of the public in India. Cambridge University Press, Cambridge

Scheufele DA, Tewksbury D (2007) Framing, agenda setting, and priming: the evolution of three media effects models. J Commun 57:9–20

Tiwari M, Pandey R (2013) BattleGround UP: politics in the land of Ram. Tranquebar, Chennai

Wu HD, Coleman R (2009) Advancing agenda-setting theory: the comparative strength and new contingent conditions of the two levels of agenda setting effects. J Mass Commun Q 86(4) (winter)

Chapter 4
From Regional to National: The Making of Uttar Pradesh as the Heartland of Indian Politics in Media

Abstract During Uttar Pradesh election campaign myriad political discourses surfaced across the mediascapes in different political interpretations and representations. Mediascapes discussed these images (news) as spatial imageries to understand the importance of heartland politics in the country. During election campaign mediascapes represented Uttar Pradesh as regional space along with politics on heartland, politics on political manifesto, politics on ECI, politics of smaller states or bifurcation of the state into four new states and politics on statues.

Keywords Regional · National · Exit poll · Election commission of India (ECI) · Cultural capital

When news appears in the media, media becomes a medium that disseminates the same information for specific purposes to their target audiences. Such news can then be understood as being mediated. Being mediated shows/represents the cartography of the space in a mediated form and an individual gathers his/her impression of a particular space/place, from media sources. The mediated cartography[1] shows the information in representations. This information might be political, cultural, social, economical, gender, religious, and geographical and can even vary from subject to subject or image to image. Sonia Livingstone has examined the media effects in two ways and argues, "First, the media mediate everything, entering into and shaping the mundane yet significant relations among individuals and between individuals and society. Second *as result* that the media mediate more than ever before, ushering historical changes that may be judged more or less critically. The first is a claim about meaning, process, connection while second is a claim about the gradual transformation of power relations among institutions and

[1]For the word "mediated," refer to the discussion in the preceding chapters. The word cartography is about making maps. Noted cartography historian Harley is known for his paper on *deconstructing the map*. He draws inspiration from the work of Derrida and Foucault to situate maps as players in a system of power/knowledge, that were not autonomous from society but rather believed that maps can never be "above the politics of knowledge" (quote in Crampton, 2010: 85). Here, the meaning of cartography is quite political where media is supposed to draw a contour of politics and shaping and reshaping the space accordingly.

S. Moinuddin, *Mediascape and The State*, Springer Geography,
DOI 10.1007/978-3-319-51932-6_4

publics; these claims are, of course, mutually interdependent" [quote in Lundby 2009, see in forward chapter (xi)]. This study discusses political discourses in the form of *news* portrayed in media during the 2012 Assembly election campaign.

The politics of Uttar Pradesh have been discussed widely in both formal and informal manner in the media. It is not because Uttar Pradesh is an important political state but how the politics of Uttar Pradesh is supposed to have shaped the future of UPA and national politics after the 2012 Assembly elections. Media brings its own perspectives while covering any issue either for business interests or political affiliations. Along with Uttar Pradesh, four more states went for Assembly elections but none of the states were covered by the media as compared to the politics of Uttar Pradesh. Media covered almost all major political discourses during the election campaign in the state. These political discourses shaped some of the political images (issues) and these images were discussed in the media with from various political perspectives. These images became an intervening junction in the media and addressed the spatiality of Uttar Pradesh with reference to politics, development, caste, religion, FDI retail, backwardness, corruption, divisions of Uttar Pradesh, statues of Elephants and the anti-incumbency move against Mayawati among others issues were discussed between the national and regional political parties, where media did an incredible job in bringing the political discourse into the public domain.

Every media had its own way of representing the issues. *The Times of India* started a special page to look into the election issues and named the page the "Dance of Democracy" and ran the page till the elections were over. The page was exclusively designed to cover some of the most prominent and important political debates and discussions that were discussed during the elections. Meanwhile, TV news channels held exclusive discussions during prime time to cover these issues.

This chapter discusses three political images pertaining to the state government in the present political context. Hence, the political news was categorized in order to locate Uttar Pradesh politics as raised and represented by the media. The issues are multi-dimensional in nature and map the politics of the state. The issues reflect a broader image with a national perspective, where the role of Uttar Pradesh is evident. The issues are local/regional in nature but have a national appeal or carry political potential to change the politics in the country and the state as well. The issues are centred on the BSP government or Mayawati and it is considered that the role of BSP government was pivotal in shaping these issues and giving them a national character. The three images were—"Uttar Pradesh: the nerve of politics; Division of Uttar Pradesh and; Statues, Elephant, Parks and Memorials."

4.1 Representations and Productions of the Mediated Heartland

Uttar Pradesh is the most politically regarded state as it has given the maximum number of Prime Ministers to the nation. There is an understanding across political parties that those who rule in the state will rule over the Central Government or can

Image 4.1 Election commission of India. *Source* Aaj Tak, December 24, 2011

influence the Central Government as well. The Uttar Pradesh Assembly elections of 2012 became a political platform when political discourses were discussed wherein the role of media was clearly evident. This section discusses the initial public and political discourses appearing in the media and sees them as being the preparatory stages of an election. These discourses were interpreted in different ways–both by the media and political parties and the electorate the common people on the other. Prior to the announcement of the election notification, informally the agenda and issue of the election were very much in the political and mediaspace. The election process starts with the announcement of election notification by the Election Commission of India (ECI) (Image 4.1). However, the political rally, the election manifesto, the political campaign and exit poll are considered as the stages of an election. How have these stages appeared in the media over time? These stages are considered as being the initial phases towards the construction of the state of Uttar Pradesh as the "heartland".

The political parties welcomed the decision of ECI to hold Assembly elections in 2012—with media covered and statements of political leaders and particulars and *welcome the decision of ECI for seven phase Assembly election.*[2] The BJP President, Rajnath Singh said that there would have been no need to hold a seven-phase election, if the law and order situation in Uttar Pradesh had been better and he even welcomed the decision of a seven-phase election schedule in Uttar Pradesh. However, the BJP demanded the ECI to remove the bureaucrats and police officials who were appointed during BSP's regime with immediate effect and Mukhtar Abbas Naqvi said that those officials were behaving like agent and spokesperson of BSP. In light of this, ECI made some routine transfers and issued directions to officials to conduct fair and free elections. BJP welcomed the seven-phase election schedule in the state and considered the BSP's misrule as the reason behind the decision taken by the ECI. However, ECI had its own logic and argument behind such a long election schedule to conduct free and fair elections. In fact, it is the duty of ECI to ensure safe and fair elections in the country. Political parties including BJP covertly attacked the Mayawati government and alleged that BSP might sabotage the fair and free election

[2]See for details in news under headline "सात चरणों में चुनाव के एलान का स्वागत" on December 25, 2011 on Aaj Tak.

process and urged the ECI to take proper action to conduct fair elections. While Akhilesh Yadav applauded the ECI and reiterated that his *Cycle* will tour the state and ensure the defeat of BSP in the state. Abhishek Manu Singhvi of the Congress Party also said that the Congress will perform well in the state.

The announcement of the ECI in the media to hold a seven-phase election in Uttar Pradesh and single phase election in Punjab, Uttarakhand, Manipur and Goa was discussed in the media in light of geographical, political, social, cultural aspects. ECI was aware of the complexities in Uttar Pradesh and Uttar Pradesh's position as a politically strategic state and therefore, made all possible arrangements to conduct fair and free elections in the state. Media portrayed the views of all concerned political parties' view on the issue. The announcement of the election schedule is considered the start of the formal election process and political parties can hold election rallies, although they can hold a rally anytime but after the election schedule, the political parties have certain limitations and are supposed to follow the guidelines.

The political rallies are also considered important "public" gatherings where political parties are supposed to address political discourses. Political rallies are often known for two things; a helicopter and a Bollywood celebrity at the venue. However, politicians astutely use both these components (Helicopter and Bollywood celebrity) to gather and inform people near the rally venue about the leaders who are attending the rally.[3] Apart from political rallies, political discussions also gear up around the village *Chaupal-Chabutra*, *Dhaba*, *saloon* and other such public spaces where people usually discuss the political calculations and share information with each other for political purpose.[4] How do the rural people conceive the Helicopter and Bollywood celebrity? The visualization of a helicopter amongst rural people is quite fascinating and therefore people generally arrive at the rally venue to see the helicopter from a close proximity and experience the sand blowing around the venue during landing and take-off. Bollywood celebrities similarly are also found fascinating and some people travel a long distance to catch a glimpse of the film stars. Political leaders have learnt the mindset of the people and try to bring both a helicopter and a Bollywood celebrity at least once in their constituency to ensure huge crowds at the venue. People come to see these elements rather than listen to the political discussion or discourse. Leaders also use some other methods to attract people during the 2012 elections. A local music party performed a musical programme in Gopalpur constituency of Azamgarh district for a candidate of the Congress Party (Image 4.2).

---

[3]Tiwari and Pandey (2013, pp. 15–6).

[4]Read in context to, Habermas (1989). The Structural Transformation of the Public Sphere, (translated by Burger T.). Cambridge: Polity Press: 31–8, 41–3. As Habermas points out, the early public, then, comprised both court and "town." The shape of the court and town were further restructured and by the seventeenth century coffee houses, *salon* become the space where intellectuals met and discussed to legitimate itself in these coffee houses (Habermas "Institutions of Public Sphere" in Oliver Boyd-Barret and Chris Newbold (2012), (eds.). Approaches to Media: A Reader. Jaipur: Rawat Publications: 235–44).

Image 4.2 Musical campaign. *Source* The Hindu, February 10, 2012*. *Different strokes: A local music party campaigning for Gopalpur constituency Congress candidate at Azamgarh on Wednesday while (*right*) another campaigner opts for a rickshaw at Mubarkpur in Azamgarh district—Images: Subir Roy

Similarly, Priyanka Gandhi Vadra, the daughter of Congress Party President Sonia Gandhi, brought her children to the dais during a political meeting and tried to influence the crowd through sentiment (Image 4.3) appeared in the media.

Political rallies serve as one of the mediums to inform people about political issues and are considered a medium wherein people and leaders have direct links with each other and allow the space to discuss issues in close interaction. During elections, politicians use various methods to lure voters and to convey his/her political message and political parties often describe political manifestoes in detail. The political manifesto is considered a political activity and politicians prepare it before the election. During the election campaigns, every political party organizes *nukad natak* (street play) to address the actual issues with a local tone and flavour whereas in some places "*candidates hire bar girls to lure voters in Uttar Pradesh.*"[5] In an interview, a bar dancer from Mumbai, Chandni (name changed) said "Candidates book us for dance shows to lure voters and keep the morale of party workers high." Chandni is not alone. One of the co-ordinators of such bar dancers' group from Mumbai said "In the past few days, there has been a huge increase in calls being made to us by the clients (read politicians), who want bar dancers to entertain gatherings at their shows." "We have been provided good accommodation besides other perks," he added. Bar dancers are in demand all over the state. It is mostly small-time leaders who are splurging to make a killing in the polls," he said. Shiva, who arranges these shows, said "I have 23 dancers, including local girls and

[5]See in detail in news under headline "Candidates hire bar girls to lure voters in Uttar Pradesh" on January 18, 2012 on TOI, http://timesofindia.indiatimes.com/city/kanpur/candidates-hire-bar-girls-to-lure-voters-in-up/articleshow/11533058.cms, accessed January 18, 2012.

Image 4.3 Politics of legacy. *Source* The Hindu, February 10, 2012*. *Priyanka Vadra with her children at a Congress election rally in Rasulpur, Amethi, on Thursday—IMAGE: PTI

few from Mumbai. They are booked as of now by political candidates. I have diaries with details of advance amounts paid by political parties keen on such shows."[6]

4.2 Mapping the Political Manifesto Through Media

The political manifesto is usually seen as a detachable part of politics and is prepared to establish an issue-based communication with voters. Every political party is supposed to make a political manifesto. Therefore, every political party generally prepares a manifesto to make voters aware about their governance plan, if they assume power. The content of a manifesto varies from one political party to another and sometimes political parties even announce some freebies in their manifesto. During the Uttar Pradesh Assembly elections, almost all political parties launched some version or the other of mediated information in the manifesto to reach out to the people and were widely circulated in the media. Rahul Gandhi trying hard to build relationship with people and therefore arrived in Allahabad to celebrate the birth anniversary of Jawaharlal Nehru in order to remind the people of his family's ancestral relationship with them (Image 4.4). The *Congress Party launched their theme songs* "Utho, Jago, Badlo, Uttar Pradesh…" with the title song being. The lyrics of the song is almost like a message from Rahul Gandhi for the people of Uttar Pradesh.[7]

---

[6]See in detail in news under headline "Candidates hire bar girls to lure voters in Uttar Pradesh" on January 18, 2012 on TOI, http://timesofindia.indiatimes.com/city/kanpur/candidates-hire-bar-girls-to-lure-voters-in-up/articleshow/11533058.cms, accessed January 18, 2012.

[7]See details in news, under the headline "यूपी के लिए कांग्रेस का थीम सॉन्ग लॉन्च" on January 5, 2012 on Aaj Tak. The theme song was "*Utho, Jago, Badlo, Uttar Pradesh*" (Get Up, Wake Up, Change Uttar Pradesh). However, some controversy lures the theme songs and quite similar to that of A.R. Rahman's "Jai Ho," but the Congress Party declined any charge of copyrights over the songs. The full lyrics was as;

Utho, Jaago, Badlo, Uttar Pradesh
(Get Up, Wake Up, Change Uttar Pradesh!)

Image 4.4 Media portrayed Rahul Gandhi as trying hard. *Source* The Hindu, November 15, 2011 (Rahul Gandhi interacting with people at Anand Bhavan, Allahabad—he arrived to celebrate the birth anniversary of Jawaharlal Nehru)

The theme songs of all political parties could be heard during the election campaign. Media coverage depicted Rahul Gandhi promoting the songs for the election campaign which were being played while he was sitting on the roof of the vehicle and travelling across the state (Image 4.5).

In addition, the Congress Party also launched a "Vision Document 2020" was widely concerned by the media, to describe the party's roadmap for Uttar Pradesh's development and ensured the participation of every community in the development process in the state. The main theme of the document was, "Justice, Development and Right" (*Nyaya, Vikas aur Adhikar*), and the document was released in ten important cities in the state.[8] The document included plans to ensure everyone's participation in the development and making of the state. This included some ambitious plans like the setting up of 1000 skill development centres which will provide jobs to 20 lakh unemployed youths, plans to check the persecution against females in the state, special development plans for regions like Bundelkhand.[9] Meanwhile, SP and BJP announced in media that the distribution of laptops and tablets amongst students, when they would come to power, however, *SP will give*

(Footnote 7 continued)

Aaja Aaja U.P. Ke Vikas Abhiyan Ke Liye; Aaja Aja Rahul Ji Ke Pegam Ke Liye
(Come together for U.P.'s progress; come together for Rahul [Gandhi's] vision)
Rahul Ji Ki Soch Se Hum U.P. Ko Badhayenge; Mehnatash Ko Adhikar Dilayenge
(Rahul's perspective will change Uttar Pradesh; assure the worthy of their rights)
Nojavan Dar Dar Ki Thokar Na Khayega; Apne Hi U.P. Mein Rozgar Payega'

(The youth will not succumb to fear under our leadership; they will attain employment in their own state, U.P.) http://blogs.wsj.com/indiarealtime/2012/02/09/congress-remakes-jai-ho-into-party-anthem/, accessed February 9, 2012.

[8]The cities were Lucknow, Kanpur, Varanasi, Allahabad, Gorakhpur, Jhansi, Agra, Meerut, Shahjahanpur and Moradabad where Congress party's "Vision Document 2020" launched, http://webcache.googleusercontent.com/search?q=cache:http://indiatoday.intoday.in/story/congress-releases-vision-document-for-uttar-pradesh/1/170813.html, accessed February 9, 2012.

[9]See in details in news under the headline "Uttar Pradesh Polls: Cong vision document up ahead" on January 10, 2012 on CNN IBN, http://ibnlive.in.com/news/up-polls-congress-vision-document-up-ahead/219661-37-64.html?utm_source=ref_article219661-37-64.html?utm_source=ref_article, accessed January 10, 2012.

Image 4.5 Mode of election campaign. *Source* Aaj Tak, January 5, 2012

laptops to student.[10] They promised to give laptops to the students who passed Class XII and tablets to those who passed Class X. SP manifesto further promised about "unemployment compensation," "free education to girls" and "pension plans for old age peoples." BJP launched its manifesto in the media with title "*Hamare Sapnon ka Uttar Pradesh*" (Uttar Pradesh of our dream) where BJP reconciled the "*swadeshi*" (indigenous) with "*ethnic*" authenticity.[11] Hence, *BJP manifesto: promised for laptops and tablets*[12] and further they promise that *they will build Ram temple, if they came to power*[13] and stop 4.5% reservation provision to minorities under OBC quota and provide 50% reservation to women in local administration. However, BSP claimed that they did not believe in a political manifesto and believed that their hard work would win people's confidence. Thus, every political party had its own vision's for the state.

The interesting part regarding the manifesto politics was that SP never supported information technology and English education; rather the 2012 Assembly elections were noted as a marking shift when SP announced the giving away of laptops and tablets to the students. The shift was considered a phenomenal turn in SP's policy. But such a shift was possible because of Akhilesh Yadav, who was looking after the election affairs and believed in modern technology. Later, he went on to become the youngest Chief Minister of the state. These phenomenal shifts widely circulated and reported in media and played an instrumental role and shaped the spatial politics and thus, SP emerged as the winner in the elections. Apart from this shift of course, there were other issues that shaped the electoral politics in favour of SP.

While the BJP also promised laptops and tablets to students, they also tried to polarize society after raising the Ram temple issue. But this time, the issue did not

[10]See for details in news under the headline "सपा देगी छात्रों को मुफ्त में लैपटॉप" on 21.1.2012 on Zee News.

[11]Prahladan (2012). "Uttar Pradesh Assembly Elections: Politics of 'Belonging' or 'Belongings'?". EPW. Vol. XLVII No 6. February 11, Mumbai. pp. 10–11.

[12]See for details in news under the headline "बीजेपी का घोषणापत्र: लैपटॉप, टैबलेट का वादा" on January 27, 2012 on Zee News.

[13]See for details in news under the headline "बीजेपी का घोषणापत्र: लैपटॉप, टैबलेट का वादा" on January 27, 2012 on Zee News.

Image 4.6 Money does matter. *Source* TOI, January 18, 2012. p. 14 (dance of democracy)

attract people. However, all political parties criticized BJP's manifesto for raising the controversial discourse again. Hence, during elections, it is odd to think that the manifesto occupies prime position; rather money becomes one of the fundamental means for political parties to gain an edge over their competitors. Media often apprises the people regarding the way that money power works during elections and the fact that more or less every political party indulges in such behaviour. In the caricature (Image 4.6) where leaders riding in a jeep through the streets with peoples lined up and leaders throwing money (currency notes) in form of paper plane towards the peoples and asking "vote us back".

It is believed that a manifesto is gradually losing its glory and is being replaced by "*moneyfesto*" (money), as the above cartoon depicts. The way in which these unethical means are supported has created an alarming situation regarding the future of our democratic practices in the country. Media portrayed the politics in a contentious manner where politicians were shown in a negative way rather than the politicians focussing on developmental issues.

4.3 Political Campaigns and the Representation of Uttar Pradesh in Media

During political campaigns, the media becomes a pivotal medium to follow the nature of politics and political debates. The political discussion touches almost all the relevant issues that surfaced in the elections. The most prominent issues that were discussed in the media during 2012 elections were caste and religion, where every political party made a point to clarify their political party's stand about the issue. Meanwhile, both these issues are considered the two most important political characteristics of state politics, which to some extent shapes the national politics making Uttar Pradesh the "heartland" of Indian politics. Media portrayed these issues in order to understand the ways in which politicians were mapping the issue around the "heartland" and portrayed caste and religion as a viable political option to assume power in the state. Media has positive impact by showing the problems and helping in it immediately."[14]

4.4 Representing Caste and Communal Politics as the Core of Uttar Pradesh Politics

In Uttar Pradesh politics, caste and religion does matter and is visible in (Fig. 4.1) in the figure discussed. In the Assembly election 2012, Congress had given maximum election seats to the Dalits in comparison to other parties in the state. BJP had given maximum tickets to the OBCs in comparison to other parties while BSP had given maximum seats to the OBCs. BSP had given 85 seats to the Muslims while Congress Party and SP had given 70 and 80 respectively in the election. For example, BSP experimented with the caste configuration in the form of "social engineering" when she brought Brahmin and Dalit castes under the same platform and SP followed "*Muslim-Yadav*" (M-Y) formation. While both BJP and Congress Party brought forward class along with marginal people in the formation. Clement Six infers that "the subject of the relationship between religion and politics in India will, therefore, also have to deal with the specificities of India as a democracy since these determine the character of political conflicts and the contest for power."[15]

However, national party including Congress Party also shaped the caste politics of Uttar Pradesh which is depicted in media in the following headlines—"*Congress digs into past to woo Brahmins ahead of U.P. election*"[16] and formed committees to look after the issue very seriously in order to associate Brahmin with Congress Party. The caste politics is often visible and The Times of India analyzed the caste politics in the state and found that "*BJP banks on Atal Bihari Vajpayee' power for Uttar Pradesh polls*"[17] to "*bait Brahmins*" and the mission resurrection of Vajpayee

[14]Interview with Anupam Shukla. He is a college student and living in Delhi for the past one year.

[15]Six (2010).

[16]See in detail in news under the headline, "Congress digs into past to woo Brahmins ahead of U.P. elections" on November 14, 2011 on The Hindu, http://www.thehindu.com/news/national/congress-digs-into-past-to-woo-brahmins-ahead-of-up-elections/article2627756.ece, accessed November 14, 2011.

[17]See in detail in news under the headline, "BJP banks on Atal Bihari Vajpayee' power for Uttar Pradesh polls" on November 6, 2011 on TOI, http://timesofindia.indiatimes.com/city/lucknow/BJP-banks-on-Atal-power-for-UP-polls/articleshow/10625905.cms? accessed November 6, 2011.

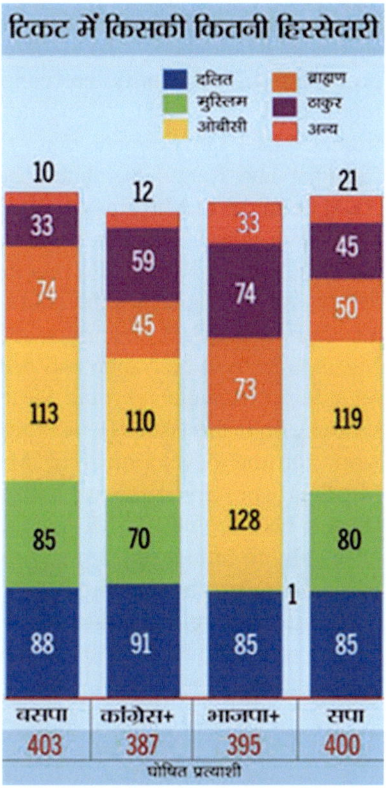

Fig. 4.1 Caste-wise ticket distributions. *Source* Aaj Tak, January 21, 2012

commenced in June, 2011 when Nitin Gadkari played glowing tributes to the leader in the national executive meeting and declared how the BJP has resolved to following the path shown by "Adarniya Atalji" (respected Atal ji).[18] Hence, both the Congress Party and the BJP also were involved in caste politics in Uttar Pradesh.

On the other hand, the BSP argued in media for *reservation for the poor forward caste people*[19] and Mayawati was quoted in the media as saying that she had written so many times to the Prime Minister of India regarding reservation for the forward castes. In fact media widely reported that. "*Mayawati describes Brahmins as* "*exploited lot*"[20] by others during Congress and BJP regime in the state. She further

[18]See in detail in news under the headline "BJP to bait Brahmins with Vajpayee" on December 21, 2011 on TOI, http://timesofindia.indiatimes.com/india/UP-polls-BJP-to-bait-Brahmins-with-Vajpayee/articleshow/11187395.cms? accessed December 12, 2011.

[19]See for detail in news under the headline "ऊंची जाति के गरीबों को मिले आरक्षण" on November 13, 2011 on Zee News.

[20]See in detail in news under the headline 'Mayawati describes Brahmins as "exploited lot" on November 14, 2011 on The Hindu, http://www.thehindu.com/news/national/mayawati-describes-brahmins-as-exploited-lot/article2624905.ece, accessed November 14, 2011.

added, "If the Congress and the BJP succeed in their game-plan, no benefit would accrue to Brahmins."[21] To some extent, the Congress tried to pacify Muslims by promising 4.5% minority reservation within the OBC quota. Mulayam, in a rally as reported in media that, "Cong not serious about Muslim quota"[22] and requested from Central Government that "the recommendations of these two committees (Sachhar and Rangnath) should be immediately implemented so that reservation could be given to Muslims. We (SP) are in its favour and have been demanding the same for a long time." His comments came in the backdrop of the then Union law minister Salman Khurshid's statement that the Centre was studying various models for providing quota within quota for the Muslim community. Mulayam Singh Yadav in a political meeting said that he had raised the matter of reservation for Muslims in the Lok Sabha and had written a letter to the Prime Minister of India, regarding the matter as well. He further alleged that the Congress Party was not serious about the issue. He was reporting as saying "our party has always tried to work for the development of Muslim community. During our regime, 14.5% Muslims were given jobs in police. The percentage has come down to 1.5 in the present BSP regime."[23] Meanwhile, BJP in the media was seen as organizing an OBC rally to influence them because BJP knew that OBCs share of population and particularly in some constituencies can play a crucial role (Image 4.7).

However, religion remains to be considered a politically contentious issue in the state as well. The Batla House encounter[24] resurfaced during political campaign that lead widespread agitations, effigy burning and demand for judicial enquiry.[25] This further politicized the issue with the BJP demanding clarification from the

[21]See in detail in news under headline 'Mayawati describes Brahmins as "exploited lot" on November 14, 2011 on The Hindu, http://www.thehindu.com/news/national/mayawati-describes-brahmins-as-exploited-lot/article2624905.ece, accessed November 14, 2011.

[22]See in detail in news under headline "Cong not serious about Muslim quota" on November 2, 2011 on TOI, http://timesofindia.indiatimes.com/city/lucknow/Cong-not-serious-about-Muslim-quota-Mulayam/articleshow/10574946.cms?, accessed November 2, 2011.

[23]See in detail in news under headline "Cong not serious about Muslim quota" on November 2, 2011 on TOI, http://timesofindia.indiatimes.com/city/lucknow/Cong-not-serious-about-Muslim-quota-Mulayam/articleshow/10574946.cms?, accessed November 2, 2011.

[24]Batla House encounter held on September 19, 2008 and officially known as operation Batla House took place in Batla House nearby the locality in Jamia Nagar, New Delhi, in which two suspected terrorists of Indian Mujaheeden (IM) Atif Amin and Mohamed Sajid were killed while two other suspects Mohd Saif and Zeeshan were arrested, while one accused Ariz Khan managed to escape. Delhi Police encounter specialist Mohan Chand Sharma, who led the police action was also killed during the incident http://en.wikipedia.org/wiki/Batla_House_encounter_case, accessed November 12, 2011. However, the encounter was discussed in contentious ways and the encounter divided the politics in the country, although the issue made some impacts in the Uttar Pradesh as well. During Assembly election 2012, the incident once again raised in order to address politics and politicians was quite successful so far.

[25]See for details under the headline "बटला हाउसकांड की न्यायिक जांच की मांग पर प्रदर्शन, पुतला फूंका" on January 12, 2012 on Rashtriya Sahara.

Image 4.7 BJP's Love for OBCs. *Source* Aaj Tak, February 12, 2012 (BJP leaders on dias)

Congress-led Central Government. This prompted the then Union Home Minister P. Chidambaram to respond and clarify the issue. It is interesting to note that the Batla House encounter occurred in Delhi and yet became a political issue of prime importance that was thought by many in the media to influence the votes in Uttar Pradesh. This perception strengthens my argument that Uttar Pradesh shares an interesting relationship with national politics as evident from the fact that issues at the regional/local level of Uttar Pradesh assume prime importance while national questions become importance concerns for the people of Uttar Pradesh.

4.5 Between Caste and Communal Politics: The Development Agenda

While both the Congress Party and BSP were in power in the Centre and in the state respectively, they shaped their political discourse around the notion of a "heartland." Mayawati, in response to the speeches made by Rahul Gandhi in media said that *Uttar Pradesh become bankrupt, if Congress formed government*[26] and acknowledged the apathy of the people during the Congress Party regime and dissuaded the voters from voting the Congress Party. The Congress Party spokesperson Abhishek Manu Singhvi, criticized Mayawati on the Mahatama Gandhi Rural Employment Guarantee Scheme (MNREGA) issue and said to media that *Mayawati does not have work but has excuses*,[27] Digvijay Singh in a rally at Siddarthnagar further alleged that the grant of one thousand million that was given by the Central Government for National Rural Health Mission (NRHM) development in Uttar Pradesh, was mismanaged. While he told media reporters that he

[26]See for details in news under headline 'कांग्रेस सरकार बनी तो यूपी हो जाएगा कंगाल' on November 28, 2011 on Rashtriya Sahara.

[27]See for details in news under the headline "मायावती के पास काम कम, बहानेबाजी ज़्यादा" on October 29, 2011 on Aaj Tak.

Image 4.8 Political training. *Source* TOI, January 21, 2012 (Digvijaya giving training Rahul-how to hit multiple target from one *arrow*)

came to see the status of yellow fever in the region (Eastern Uttar Pradesh) and said that as early as possible the Central Government would ensure the eradication programme for yellow fever in the district.

During the Uttar Pradesh Assembly elections 2012, Digvijay Singh was in the limelight as he was in-charge of the Congress working groups and coordinating with the state leaders. Digvijay Singh was a former Chief Minister of Madhya Pradesh and a senior leader in the Congress Party. The media reported several times that he was virtually guiding Rahul Gandhi during the election for a larger political role. While he refutes such charges and said that he was assisting him, being a party leader. The media nonetheless implied otherwise as evident from the caricature where he is showing Rahul how to shoot three leaders of different political parties by a single shot (Image 4.8). Interestingly, the media depicts Uttar Pradesh as the ground for training for larger political role at the national level. Thus, affirming the argument made that the ability to understand Uttar Pradesh politics will better enable one to handle national politics making Uttar Pradesh the core and as argued the heartland of Indian politics and polity.

Mayawati the then CM in her political rally at Kannauj was quoted in the media as saying *Centre was as doing step motherly with Uttar Pradesh*[28] and the Congress-led UPA did not provide financial assistance, as she (Mayawati) had been demanding for some time. She further alleged that the existence of corruption and unemployment is because of the long regime of the Congress Party in the state. And in another rally at Lucknow she made personal attacks on Rahul Gandhi and made headlines that carried following narrative: "Maya plays foreign card to target

[28]See for details in news under the headline "केंद्र ने कयिा उत्तर प्रदेश के साथ सौतेला व्यवहार" on February 13, 2012 on Aaj Tak.

Image 4.9 Contentious politics on FDI. *Source* Aaj Tak, November 25, 2011 (Uma Bharti with Anand Sharma)

Rahul & retail"[29] and condemned the Centre's decision to allow 51% FDI in multi-brand retail and said "My party will oppose the move in Parliament and outside." While she further added, "As the move (to allow FDI in retail) is aimed at benefitting Rahul Gandhi's foreign cronies, any reluctance on UPA's part to roll back the proposal would only strengthen my resolve to take some tough decisions."[30] Meanwhile the tussle between BJP and the Congress Party over the FDI issue became fierce when Uma Bharti openly opposed the FDI bill and said she will set fire to the FDI retail store. However, the BJP distanced itself from Uma Bharti's statement and said that it was her personal statement and the BJP had nothing to do with it (Image 4.9).

BSP on the other hand, vowed to never allow any foreign firm to set up base in Uttar Pradesh as long as she remained in power. The attack was a result of Rahul Gandhi's five-day visit when he directly attacked the BSP-led government and accuses them of "*graft and lawlessness*." Mayawati further added him as widely quoted in media. "He was brought up and educated abroad and his style of living and thinking is also more that of a foreigner…. He is in the habit of dashing off to foreign countries after spending a few weeks in India." "During his Uttar Pradesh sojourn, Rahul planned fun trips for his foreign companions and promoted "poverty tourism." He takes his guests to poor people's homes for entertainment and makes a mockery of their suffering," Mayawati claimed.[31] In addition she said in media "Congress is so worried over its weak and pitiful position in the state that its "prince" has to do different types of theatrics in the state everyday leaving Parliament in Delhi." She said it seemed that the Congress was getting "nightmares" about BSP's election symbol "elephant." Mayawati said in the rally, "Our

[29]See in detail in news under the headline "Maya plays foreign card to target Rahul & retail" on November 27, 2011 on TOI, http://timesofindia.indiatimes.com/city/lucknow/Maya-plays-foreign-card-to-target-Rahul-retail/articleshow/10887013.cms?, accessed November 27, 2011.

[30]See in detail in news under the headline "Maya plays foreign card to target Rahul & retail" on November 27, 2011 on TOI, http://timesofindia.indiatimes.com/city/lucknow/Maya-plays-foreign-card-to-target-Rahul-retail/articleshow/10887013.cms?, accessed November 27, 2011.

[31]See in detail in news under the headline "Maya plays foreign card to target Rahul & retail" on November 27, 2011 on TOI, http://timesofindia.indiatimes.com/city/lucknow/Maya-plays-foreign-card-to-target-Rahul-retail/articleshow/10887013.cms?, accessed November 27, 2011.

party's policies and growing influence across the country have made life difficult for Congress Party. In today's scenario, it seems that they have nightmares about our election symbol 'elephant' driving them away."[32]

At Jaunpur, according to media reports Mayawati said the *Congress should not preach about growth*[33] as in their four decades of governance, they provided opportunities to the opposition parties to rule over the state. At Raebareli she said that *Gandhi family is responsible for Raebareli's backwardness*[34] and they never looked back at the constituency after winning the election. In a public rally at Partapgarh, she attacked the Congress Party-led UPA government and said "Centre creating barriers in Uttar Pradesh's development"[35] and further alleged that the Centre was creating hindrances in the development of Uttar Pradesh. "The Congress did not want to see development in the state. That is why it did not accept the proposal for division of the state into four parts," she said. Addressing a string of election meetings, the BSP supremo asserted that if anyone tried to disrupt the free and fair poll in the state, they would be dealt with sternly. She further added that voters should exercise their franchise fearlessly and said "I will not allow *goonda* and other elements to create problems during the election. I will get them hanged upside down."[36] She cautioned people about other political parties and asked them to ensure that the BSP wins and forms the government for inclusive development in the state. She said after the Uttar Pradesh Assembly election-2012, the Congress Party will lose the *leftover* people's mandate.

Syed Sahnawaj Hussain, the BJP spokesperson, criticized the Congress Party and Rahul Gandhi for "selling big dreams to the people of Uttar Pradesh"[37] and he further attacked on Congress and said that mere speeches were not going to remove corruption. Similarly, the condition in Uttar Pradesh was not going to change with speeches. He illustrated a few examples to substantiate his arguments and said that

---

[32]See in detail in news under the headline "Rahul indulging in 'theatrics' during Uttar Pradesh campaign" on November 27, 2011 on TOI, http://timesofindia.indiatimes.com/india/Rahul-indulging-in-theatrics-during-UP-campaign-says-Mayawati/articleshow/10891698.cms?, accessed November 27, 2011.

[33]See in details in news under the headline "विकास पर कांग्रेस न बोले" on February 11, 2012 on Rashtriya Sahara.

[34]See for details in news under headline "रायबरेली के पिछड़ेपन को गांधी परिवार जिम्मेदार" on February 13, 2012 on Rashtriya Sahara.

[35]See in details in news under the headline "Centre creating barriers in Uttar Pradesh's development: Mayawati" on February 15, 2012 on CNN IBN, http://ibnlive.in.com/news/centre-creating-barriers-in-ups-development-mayawati/230570-37-73.html?utm_source=ref_article, acc essed February 15, 2012.

[36]See in details in news under the headline "Centre creating barriers in Uttar Pradesh's development: Mayawati" on February 15, 2012 on CNN IBN, http://ibnlive.in.com/news/centre-creating-barriers-in-ups-development-mayawati/230570-37-73.html?utm_source=ref_article, acc essed February 15, 2012.

[37]See in details in news under the headline "Rahul selling big dreams to peoples in Uttar Pradesh" on November 15, 2011 on TOI, http://timesofindia.indiatimes.com/india/Rahul-Gandhi-selling-big-dreams-to-people-of-UP-BJP/articleshow/10733258.cms?, accessed November 15, 2011.

Congress Party had long back went ahead with the slogan of *remove poverty* ("Garibi Hatao") but the number of people those living below the poverty line had actually gone up during the Congress rule. Syed Shanawaj Hussain adds examples of Congress regime as "the price of petro products has been raised 24 times by this government. Sugar is selling at Rs. 50 while pulses cost Rs. 100 per kg." BJP has been targeting the Congress and the UPA government on the issues of corruption and price rise in the past few months. "The money that could have helped in removing poverty was lost in the Rs. 1.76 lakh crore 2G spectrum scam. We will not sit quiet and want an answer." On atrocities against the Dalits, BJP said that Rahul Gandhi was silent on the oppression of Dalits in Congress-ruled Haryana and the minority community in Gopalgarh, Rajasthan. "Despite atrocities against the minority community in Gopalgarh, the Rajasthan Chief Minister and home minister continue in office," insisting that Congress, SP and BSP were allies at the Centre and were merely engaged in a "mock fight" in Uttar Pradesh.

"The Bharatiya Janata Party leader Rajnath Singh said Prime Minister Manmohan Singh should resign on moral grounds following the Supreme Court verdict on the 2G spectrum scam."[38] Singh addressed the public meetings in the three assembly constituencies of the district. The BJP leader said the Bahujan Samaj Party and Samajwadi Party are the parties of castes. Rajnath said if the BJP comes to power, then Mayawati and Mulayam would land in jail."[39] However, Senior BJP leader L.K. Advani in a rally at Lucknow, asked the people to compare the NDA regime with UPA I and II regime on a corruption scale while his party promised clean governance—"BJP promises graft-free govt."[40]

Mulayam Sigh Yadav aggressively attacked BSP and promised to the people that "Loans to farmers will be waived"[41] and added that "If the SP is voted to power in Uttar Pradesh, we will provide free electricity to weavers and farmers besides waiving all loans of "kisans" (farmers) up to Rs. 50,000."[42] Hence, Azam Khan

[38]See in details in news under headline "Jail for Maya, MSY if BJP comes to power: Rajnath" on February 4, 2012 on TOI, http://timesofindia.indiatimes.com/city/allahabad/jail-for-maya-msy-if-bjp-comes-to-power-rajnath/articleshow/11749466.cms, accessed February 4, 2012.

[39]See in details in news under the headline "Jail for Maya, MSY if BJP comes to power: Rajnath" on February 4, 2012 on TOI, http://timesofindia.indiatimes.com/city/allahabad/jail-for-maya-msy-if-bjp-comes-to-power-rajnath/articleshow/11749466.cms, accessed February. 4, 2012.

[40]See in detail in news under the headline "BJP promises graft free govt." on February 13, 2012 on TOI, http://timesofindia.indiatimes.com/india/BJP-promises-graft-free-govt/articleshow/11867336.cms?, accessed February 13, 2012.

[41]See in details in news under the headline "Loans to farmers will be waived" on January 17, 2012 on TOI, http://timesofindia.indiatimes.com/city/lucknow/Loans-to-farmers-will-be-waived-MSY/articleshow/11519363.cms?, accessed January 17, 2012.

[42]See in details in news under the headline "Loans to farmers will be waived" on January 17, 2012 on TOI, http://timesofindia.indiatimes.com/city/lucknow/Loans-to-farmers-will-be-waived-MSY/articleshow/11519363.cms?, accessed January 17, 2012.

said in Kausambi that *remove BSP then development would come*[43] and accuses the BSP government of mismanaging the public exchequer fund.

Interestingly, the focus of the political campaigns across party lines has been issues that will subsequently have greater impact on national politics. For example, development has engaged as the key focus not just for the national parties like Congress or BJP, which one may argue are framing their political campaign with an eye to the forthcoming parliamentary elections of 2014. Even the regional parties like SP and BSP campaigns emphasized on the development of Uttar Pradesh moving away from earlier politics of caste and religion. In fact the media coverage of the political campaigns actually brought out the probable political strategies and campaigns for the national politics and as seen in the current Lok Sabha Election 2014, where developmental politics played a key role in national politics.

4.6 Exit Poll and the Politics of Predictions

After the election was over, across the media both print- and audiovisual-conducted exit poll interpreted and predicted the political future of the State. Almost every exit poll survey predicted a hung Assembly and the SP was given apolitical edge over other political parties who were also in the running to form the government in the state (Appendix C.1). Hence, every political party was in denial mode to accept the findings of the exit poll rather they had their own assumptions about the election results. Overwhelmingly, after the elections the exit poll survey across in Uttar Pradesh media showed that *Congress appeared weak in Uttar Pradesh*.[44] Meanwhile, the Congress Party General Secretary, Janardhan Dwivedi categorically stated in the media that *Congress would sit in opposition if they did get majority*. In both opinion poll and exit poll surveys, SP was predicted as emerging as the largest political party in the Assembly. But both the surveys did not predict that the SP would gain a thumping majority rather, the surveys conducted by the media predicted that *SP can form government in Uttar Pradesh*[45] and would emerge as the largest political party in the Assembly. However, the election results showed that the SP did well in both rural and urban areas—"SP rides high in rural, urban seats"[46] and came back into power. SP got votes in all the regions of the state

[43]See for details in news under the headline "बसपा हटाओ तभी वकिास" on February 11, 2012 on Rashtriya Sahara.

[44]See for detail in news under the headline "Uttar Pradesh में मजबूत नहीं है कांग्रेस की स्थति" on March 4, 2012 on Aaj Tak.

[45]See for detail in news under the headline "यूपी में बन सकती है सपा की सरकार" on March 3, 2012 on Aaj Tak.

[46]See in details in news under the headline "SP rides high in rural, urban seats" on March 7, 2012 in TOI, http://timesofindia.indiatimes.com/city/lucknow/sp-rides-high-in-rural-urban-seats/articleshow/12170478.cms, accessed March 7, 2012.

including Bundelkhand, Poorvanchal, Awadh and Paschim Pradesh and thus, SP won the elections with a bang (Appendix C.2).

However, it was not clear who the state's next Chief Minister would be, meanwhile the youth of the party rooted for Akhilesh Yadav as reported in the media that "Youth brigade wants Akhilesh Yadav as CM."[47] Though within the SP camp it was still not clear who would become the state's next Chief Minister. Addressing his maiden post-victory press conference, Akhilesh Yadav, the man who scripted the party's historic win, did little to undo the suspense. One of his close aides was quoted in media that "We cannot ignore the fact that all the young candidates of Akhilesh ji have emerged victorious. As a democratic leader, Netaji (Mulayam Singh Yadav is also known as Netaji) is unlikely to ignore this fact."[48] Though a final decision in this regard was to be taken at the Vidhan Mandal (Legislative council) and Parliamentary board meetings, the youth candidates were hopeful. Veterans in the party said that there was a bigger role waiting in the wings for the 38-year-old MP from Kannauj. "Winning Uttar Pradesh is certainly a milestone for the party. But building the party's cadre beyond Uttar Pradesh is the next agenda; a role that is reserved for Akhilesh," a party source had reportedly said to the media.[49] Though, the resounding success of mission Uttar Pradesh was not the end of the road for the Samajwadi Party. If internal sources and media coverage are to be believed, even amidst celebrations inside and outside the party's Vikramaditya Marg office (Lucknow), the cadre was readying itself for Mission 2014, the next big focus on the party's agenda. Reports appeared in the media claiming that party sources have remarked, "We are looking at expanding SP's base in Uttarakhand, Bihar, Rajasthan and Madhya Pradesh. Preparations will begin within the next six months,"[50] The SP cadre started the preparations to project the party at the national level and even they agreed that the Uttar Pradesh Assembly elections of 2012 were the semi-finals to the General elections 2014. In a CSDS survey, when respondents were asked who they would prefer as Chief Minister if the SP came to power, Mulayam Singh or his son Akhilesh, nearly half of them opted for father. Among SP voters, two-thirds preferred Mulayam Singh[51] (Table 4.1).

[47]See in details under the headline "Youth brigade wants Akhilesh Yadav as CM" on March 7, 2012 on TOI, http://timesofindia.indiatimes.com/city/lucknow/Youth-brigade-wants-Akhilesh-Yadav-as-CM/articleshow/12169590.cms?, accessed March 7, 2012.

[48]See in detail under the headline "Youth brigade wants Akhilesh Yadav as CM" on March 7, 2012 on TOI, http://timesofindia.indiatimes.com/city/lucknow/Youth-brigade-wants-Akhilesh-Yadav-as-CM/articleshow/12169590.cms?, accessed March 7, 2012.

[49]See in details under the headline "Youth brigade wants Akhilesh Yadav as CM" on March 7, 2012 on TOI, http://timesofindia.indiatimes.com/city/lucknow/Youth-brigade-wants-Akhilesh-Yadav-as-CM/articleshow/12169590.cms?, accessed March 7, 2012.

[50]See in details under the headline "Youth brigade wants Akhilesh Yadav as CM" on March 7, 2012 on TOI, http://timesofindia.indiatimes.com/city/lucknow/Youth-brigade-wants-Akhilesh-Yadav-as-CM/articleshow/12169590.cms?

[51]Kumar et al. (2012).

Table 4.1 Citizens' choice for Chief Minister if SP wins

Who should be Chief Minister if SP wins?	All	SP voters
Mulayam Singh Yadav	48	65
Akhilesh Yadav	20	16

Source CSDS, 2012 (EPW, April 7, 2012. Vol. XLVII NO 14)

"Akhilesh Yadav thanks Uttar Pradesh voters,"[52] was the headline and being the GenNext face of SP, he said in an interview that his first priority would be to fulfil all promises that his party had made during the election campaign. In his first press conference after its stunning victory, he started off by thanking the people of the state and promised to do whatever it takes to put the party's manifesto to shape on the ground. "The final tally is yet to come in but the trends are encouraging for which the party is thankful to the people of Uttar Pradesh," he said. "We will do everything possible to live up to the expectations of the people of Uttar Pradesh and not let them down for believing in the SP," he added.[53]

The election results were interpreted in various political contexts including the future shape of UPA, looming midterm elections, presidential election, union budget, general election of 2014 and role of SP and BSP in national politics, etc. The political discourses were discussed across the media in speculative ways.

4.7 The Division of Uttar Pradesh and the Making and Unmaking of the Heartland

Before the elections, the Mayawati-led BSP government decided to divide Uttar Pradesh into four new states, viz. Awadh Pradesh, Purvanchal, Bundelkhand and Paschim Pradesh. The opposition parties created an uproar when they heard about the news of division of Uttar Pradesh and vowed not to allow the division of the state. The issue was discussed across media in a political light where the opposition parties termed the issue as being politically motivated. The issue was a part of a deliberate attempt by Mayawati-led BSP to divert the direction of politics in the state.

However, there has been a constant demand for separate statehood for Paschim Pradesh, Bundelkhand, Purvanchal and Awadh Pradesh from various sections of peoples of Uttar Pradesh over the years. Every time these demands are supported by different sections of political parties amidst protests by the opposition. Of course, these demands are based on political calibration and arguments are often put forth

[52]See in details in news under headline "Akhilesh Yadav thanks Uttar Pradesh voters" on March 7, 2012 on TOI, http://timesofindia.indiatimes.com/city/lucknow/Akhilesh-Yadav-thanks-UP-voters/articleshow/12169282.cms?, accessed March 7, 2012.

[53]See in details in news under the headline "Akhilesh Yadav thanks Uttar Pradesh voters" on March 7, 2012 on TOI, http://timesofindia.indiatimes.com/city/lucknow/Akhilesh-Yadav-thanks-UP-voters/articleshow/12169282.cms?, accessed March 7, 2012.

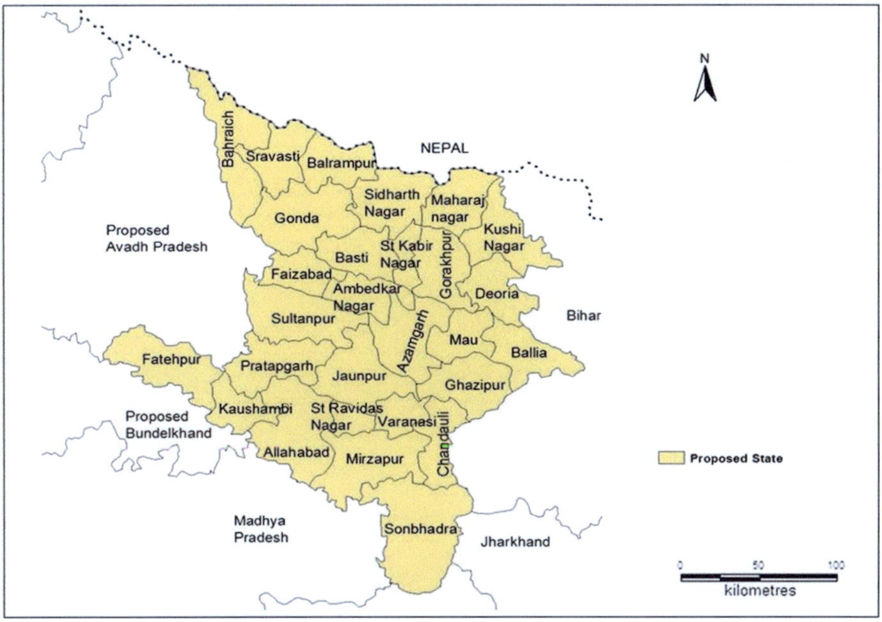

Map 4.1 Purvanchal (proposed state in Uttar Pradesh)

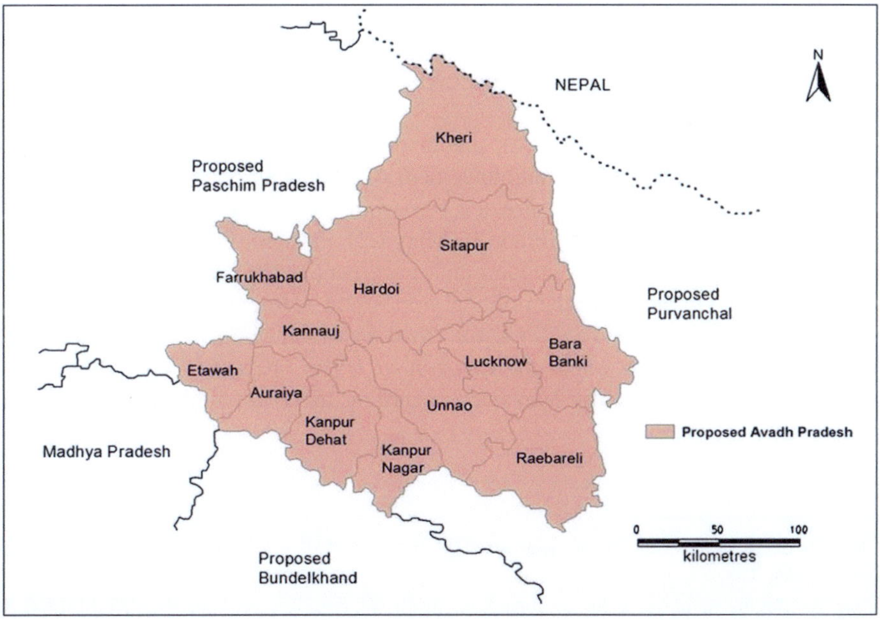

Map 4.2 Awadh Pradesh (proposed state in Uttar Pradesh)

Map 4.3 Paschim Pradesh (proposed state in Uttar Pradesh)

to rationalize their demands for separate statehood. After political deliberation, the BSP government proposed four divisions of Uttar Pradesh to fulfil people's demand. Ever since the demand for the division was made, SP and other opposition parties opposed the decision.

The issue was discussed across media during the second and third week of November, 2011. The sudden announcement of Mayawati on November 15, 2011 that *Uttar Pradesh would be divided into four parts*[54] and the states will be Purvanchal (Map 4.1), Awadh Pradesh (Map 4.2), Paschim Pradesh (Map 4.3) and Bundelkhand (Map 4.4). The sudden announcement by Mayawati was interpreted by the media as an attack on the political opposition before the elections. Mulayam Singh Yadav said that *division of Uttar Pradesh was a political conspiracy.*[55] He added that *we shall not allow the division of Uttar Pradesh.*[56] He categorically denied the decision taken by Mayawati regarding the issue while stating that it is

[54]See for detail in news under the headline “यूपी से टूटकर बनेंगे 4 अलग राज् य” on November 15, 2011 on Aaj Tak.

[55]See for details in news under the headline “राज्य को वभिाजति करने का प्रस्ताव मायावती की राजनीतकि साजशि” on November 15, 2011 on Aaj Tak.

[56]See for details in news under the headline “हम यूपी का वभिाजन नहीं होने देंगे” on November 16, 2011 on Aaj Tak.

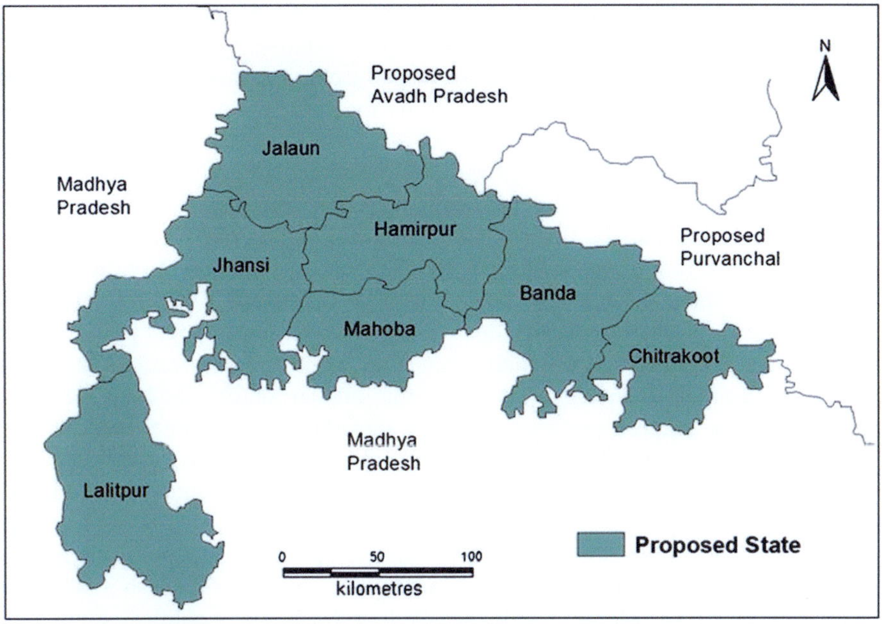

Map 4.4 Bundelkhand (proposed state in Uttar Pradesh)

merely an *election stunt* while in her response BSP said that *Uttar Pradesh division is not a political stunt*.[57]

Hence, in the media opposition questioned the timing of the move to divide Uttar Pradesh and Rajnath Singh of BJP said that BSP did nothing during their governance therefore, now, when the Assembly elections were round the corner, she was raising the issue to divert public attention from corruption and the pathetic law and order situation in the state. He further added, "As per the Constitution, the Chief Minister cannot bring any bill for division of state. At the most, she can bring a proposal in this regard in the winter session of the state Assembly but I think first to assess the feasibility of creating new states."[58] On the other hand in media, Uttar Pradesh Congress Legislature party leader Pramod Tiwari said that the Congress was not against smaller states. In fact, he added, the Congress in Uttar Pradesh had already demanded separate states of Poorvanchal and Bundelkhand earlier. "The bigger question is here that why Uttar Pradesh Chief Minister has raised the issue

[57]See for details in news under the headline "'यूपी का वभिाजन चुनावी स् टंट नहीं" on November 21, 2011 on Aaj Tak.

[58]See in details in news under the headline "Opposition questions timing of move to divide Uttar Pradesh" on TOI on November 10, 2011, http://timesofindia.indiatimes.com/city/lucknow/Opposition-questions-timing-of-move-to-divide-UP/articleshow/10674830.cms?, accessed November 10, 2011.

on the *eve* of the elections," Jan Kranti Party president Kalyan Singh, too questioned the timing of Uttar Pradesh Chief Minister's move.[59]

On the other hand, smaller political outfits seeking creation of Bundelkhand, Poorvanchal and Paschim Pradesh had come together to contest 2012 Assembly elections.

The movement led by Bundelkhand Congress, a new political outfit recently launched by actor-turned politician Raja Bundela, Purvanchal Morcha led by Shashi Kant Yadav and Harit Kranti Dal led by Ashok Sharma[60] were members of the National Federation for New States (NFNS), an organization of outfits seeking the creation of small states across the country like Telangana, Vidarbha. They were of the opinion "We tried to bring together all national and regional parties which have supported the demand of smaller states but none appeared serious in reality, hence we decided to create a platform for smaller outfits fighting for small states. The Uttar Pradesh elections will be a semi-final and the final battle will be fought in the 2014 general elections. If we are able to make even a small impact in Assembly elections, it will change the political scenario," said Bundela, who is also the co-convener of NFNS."[61]

However, the opposition parties were not happy with the proposal for the division of the state—"SP, BJP, Cong slam Maya's 'split Uttar Pradesh' call"[62] and a senior leader of SP, Mohan Singh said, "this is Mayawati's political stunt. Such action without taking into confidence all the political parties in Uttar Pradesh is wrong. We will oppose the move." Meanwhile, Mayawati's argument that was represented in media, to divide the state into four new states was interpreted on a political canvass wherein she believed that smaller states would be viable for administration and development work.[63] She demarcated the boundaries of these four new states (Map 4.5). Though the political parties rejected Mayawati's hypothesis about smaller states and termed the decision as being purely motivated by politics to hide her failures. Mayawati renamed Harit Pradesh and called it

---

[59]See in details in news under the headline "Opposition questions timing of move to divide Uttar Pradesh" on TOI on November 10, 2011, http://timesofindia.indiatimes.com/city/lucknow/Opposition-questions-timing-of-move-to-divide-UP/articleshow/10674830.cms?, accessed November 10, 2011.

[60]See in details in news under the headline "Opposition questions timing of move to divide Uttar Pradesh" on TOI on November 10, 2011, http://timesofindia.indiatimes.com/city/lucknow/Opposition-questions-timing-of-move-to-divide-UP/articleshow/10674830.cms?, accessed November 10, 2011.

[61]See in details in news under the headline "Opposition questions timing of move to divide Uttar Pradesh" on TOI on November 10, 2011, http://timesofindia.indiatimes.com/city/lucknow/Opposition-questions-timing-of-move-to-divide-UP/articleshow/10674830.cms?, accessed November 10, 2011.

[62]See in details in news under headline "SP, BJP, Cong slam Maya's 'split Uttar Pradesh' call" on CNN IBN on November 15, 2011 at 01:43 PM IST, http://ibnlive.in.com/news/sp-bjp-cong-slam-mayas-split-up-call/202547-37-66.html?utm_source=ref_article, accessed November 15, 2011.

[63]However, such demand of divide Uttar Pradesh into four smaller states was already made in the past when "Kokab Hamid from the platform of Rashtriya Lok Dal (RLD) introduced a resolution on the floor of the Uttar Pradesh legislative assembly for the division of the Uttar Pradesh into four states including the Harit Pradesh". Jagpal Singh "Politics of Harit Pradesh: The Case of Western Uttar Pradesh as a Separate State" EPW, August 4, 2001: 2961–67.

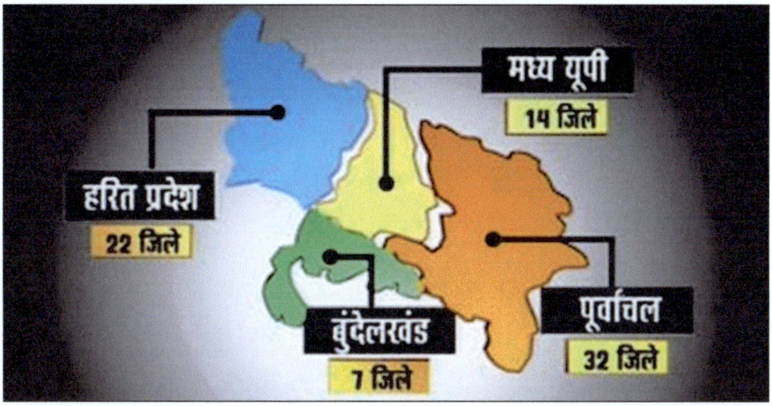

Map 4.5 Proposed division of Uttar Pradesh. Source Aaj Tak, November 15, 2011

Paschim Pradesh instead of what the RLD proposed.[64] However, "SP along with BJP opposed the RLD's resolutions in the Uttar Pradesh assembly suggesting the division of Uttar Pradesh into four more states."[65]

While BJP leader Balbir Punj was reported as saying in the media, "Mayawati's move is highly opportunist move and politically motivated. She should have first got the resolution passed in the Uttar Pradesh Assembly. Mayawati has not taken the mandate of the people before proposing division of state into 4 parts. This needs an academic exercise before making such a proposal." Congress leader Jagdambika Pal (During 2014 parliament elections he joined BJP) also said that it is a political move which is why she had written a letter to the PM and not moved a resolution in the house. He added, "If a resolution is not passed in the state Assembly and then sent to the Central Government, it cannot act on this." Another Congress leader Digvijay Singh was reported as saying, "We will take a call on supporting the resolution in the state Assembly after consulting our state leaders. Congress Party has already said that there should be a second State Reorganisation Committee to look into the demand of small states."[66]

---

[64]"Over the years different names have been given to the proposed state, e.g. Doab Pradesh, Ganna Pradesh, Kisan Pradesh, Jatistan Pradesh, Brij Pradesh, Pashchimanchal Pradesh and Harit Pradesh. Harit Pradesh, literally meaning Green province and signifying the agricultural prosperity of the region, is the latest nomenclature given to it in June 2000". See For detail. Jagpal Singh "Politics of Harit Pradesh: The Case of Western Uttar Pradesh as a Separate State" EPW, August 4, 2001: 2961–67.

[65]Ibid.: 2965.

[66]See in detail in news under the headline "SP, BJP, Cong slam Maya's split Uttar Pradesh' call" on CNN IBN on November 15, 2011 at 01:43 PM IST, http://ibnlive.in.com/news/sp-bjp-cong-slam-mayas-split-up-call/202547-37-66.html?utm_source=ref_article, accessed November 15, 2011.

Image 4.10 Sanctioning the division of Uttar Pradesh. *Source* Aaj Tak, November 26, 2011*. *Mayawati attending Legislative proceeding during passing the Uttar Pradesh split resolution

However, the headlines read—"Uttar Pradesh Assembly passes resolution to divide state into 4 parts"[67] (Image 4.10), "the resolution was passed with a voice vote and then thereafter the proceeding of the Assembly was sine die immediately after the passage of the resolution seeking to divide the state. The Assembly was adjourned for an hour amid uproar in the house and proceeding of the Assembly disturbed several times by opposition parties and they brought notice of no confidence motion in amid of protests. In between, the opposition and BSP members exchange talks over the issue of no confidence and claimed the supports of maximum legislative members in the Assembly over the issue of no confidence."[68] Meanwhile, news were being reported that, "the legislative council was deferred after opposition raised issue of BSP rebel MLC Babu Singh Kushwaha."[69]

Media portrayed Mayawati as being an instrumental leader who carved the division plan very strategically (Image 4.11). In this figure, the cartoonist has tried to express their idea about Uttar Pradesh and how she (Mayawati) is determined to divide Uttar Pradesh into four new states for her own political reasons. As depicted in the cartoon, the opposition parties are trying hard to break Mayawati's and BSP's

[67]See in detail in news under the headline "Uttar Pradesh Assembly passes resolution to divide state into 4 parts" on November 21, 2011 on TOI, http://timesofindia.indiatimes.com/india/Uttar-Pradesh-Assembly-passes-resolution-to-divide-state-into-4-parts/articleshow/10814650.cms?, accessed November 21, 2011.

[68]See in details in news under the headline "Uttar Pradesh Assembly passes resolution to divide state into 4 parts" on November 21, 2011 on TOI, http://timesofindia.indiatimes.com/india/Uttar-Pradesh-Assembly-passes-resolution-to-divide-state-into-4-parts/articleshow/10814650.cms?, accessed November 21, 2011.

[69]See in details in news under headline 'Uttar Pradesh Assembly passes resolution to divide state into 4 parts' on November 21, 2011 on TOI, http://timesofindia.indiatimes.com/india/Uttar-Pradesh-Assembly-passes-resolution-to-divide-state-into-4-parts/articleshow/10814650.cms?, accessed November 21, 2011.

Image 4.11 Political battle on division of Uttar Pradesh. *Source* TOI, November 20, 2011, p. 21

citadel. The cartoonist sees the whole episode in a political context where Mayawati is in strong position and every political party including Anna Hazare is struggling to seize Mayawati interestingly, she is shown as reluctant to divide the state into four new states. However, every political party was attacking Mayawati and criticizing her on political grounds while few smaller political parties supported the move for political reasons including RLD.

RLD reiterated the need for a separate Harit Pradesh and moved a private bill in this regard but the bill was not passed. BJP had demanded that the decision to split Uttar Pradesh should be taken after deliberation. The Party said that it supports Uttar Pradesh's division but it should not be done in a hurry. On the other hand, Congress leader Pramod Tiwari said the party will support a no-confidence motion brought by a secular party, indicating that it will not stand with BJP in the House. Pramod Tiwari also said that the resolution to reorganise Uttar Pradesh into four new states should include the demand to constitute state reorganisation commission by the Parliament for division of Uttar Pradesh. This, he said, will give the resolution a legal validity; otherwise it would be a merely "politically motivated" statement of Mayawati.

The sudden announcement of making four new states was considered as a master stroke on the part of Mayawati in media. Although she was successful in raising the issue and laid the political foundation for future politics in the state, she also demonstrated that she knew the legal provisions for creating a new state and despite

Table 4.2 Citizens' opinion on division of Uttar Pradesh

Opinion on division of Uttar Pradesh	Support to division of Uttar Pradesh		Oppose to division of Uttar Pradesh	
	2011	2012	2011	2012
Overall	29	37	32	35
Purvanchal	28	36	30	38
Awadh	29	31	36	45
Bundelkhand	20	32	22	27
Paschim	32	42	36	30

Source CSDS, 2012 (EPW, April 7, 2012. Vol. XLVII NO 14)

that she had passed a resolution and sent the bill to the Central Government for further process. She made her and BSPs political calculations after making a political assessment that her political presence is almost equal in the entire region (Bundelkhand, Paschim Pradesh, Awadh Pradesh and Purvanchal) in present day Uttar Pradesh. "On the issue of dividing Uttar Pradesh into smaller states, there is greater support for division today than it was seven months ago. While 37% of the respondents agreed with the statement that 'Uttar Pradesh should be divided into smaller states,' a slightly lesser proportion 35% disagreed. When the same question was asked during a survey conducted by CSDS in July 2011 (before the proposal by the Mayawati government to divide Uttar Pradesh into four smaller states), those in favours of dividing Uttar Pradesh was marginally lesser than those against it, across regions. The 2012 post-poll survey reveals that while most people in Poorvanchal and Awadh oppose the idea of division, in Bundelkhand and Paschim there are more people who are in support of the idea than those who are against it (Table 4.2)."[70] As Anupam Shukla argues that "being a large state Uttar Pradesh can be divided in at least two parts for better administration. Media is framing a public opinion by these informal debates."[71]

Thus, since the news regarding the division of Uttar Pradesh surfaced in the media, politicians categorically termed it as a "political stunt" and a mere political move to hide the failures of government and a strategy to divert political attention from the ground reality.

Hence the politics around the issue of the Division of Uttar Pradesh shaped the heartland as being a politically strategic state in the country. During the Uttar Pradesh Assembly elections 2012, SP got votes from all regions irrespective of rural and urban voters thus emerging as the winner. In the General elections 2014, BJP managed to make their presence felt, "What L.K. Advani's *rath yatra* couldn't achieve in the 1990s, Narendra Modi's juggernaut has done in Uttar Pradesh in

[70] Kumar et al. (2012).

[71] Interview with Anupam Shukla. Anupam Shukla pursuing his study from Delhi University and living since 10 years in Delhi while he belongs from Faizabad. He is a political activist.

2014: a near-total decimation of the entire opposition with 71 out of 80 seats and 42% vote share for BJP. It is the maximum the party has ever got. The decisive verdict, however, has a flipside: Uttar Pradesh has not sent any Muslim MP to the Lok Sabha this time. In 2009, SP had won 23, Congress 21, BSP 20 and BJP just 10."[72] Thus, media represented Uttar Pradesh as "heartland" in the national politics reiterating that widely held belief those who control Uttar Pradesh control the nation.

4.8 Statues, Elephants, Memorials and Parks: The Cultural Capital of the Heartland

During the election campaign, all political parties including SP, BJP and Congress Party criticized BSP for installing statues. The political discourse around the statues of Mayawati, elephants, memorials and parks was discussed in both public and private sphere. The opposition parties alleged that Mayawati was indulging in "caste identity politics" while BSP called the statues as "cultural capital". The political debate in media regarding the statues had begun since 2007 when Mayawati started work on them and the discussion intensified with the approaching Assembly election 2012.

The issue formally came up in the public domain when media reported the inauguration of the park in Noida-"Mayawati inaugurates Rs. 650 crore park in Noida"[73] and "Uttar Pradesh Chief Minister Mayawati on Friday inaugurated the 33-acre Dalit Prerna Sthal or the Ambedkar Park in Noida. Mayawati spent Rs. 650 crore for the development of the park, which has several statutes of Dalit leaders (Image 4.12). The park has 24 pink sandstone elephants, the electoral symbol of Mayawati's party the Bahujan Samaj Party, 12 life-size statues of Dr BR Ambedkar, Kanshi Ram and the Uttar Pradesh Chief Minister herself. The estimated cost of each statue is over Rs. 6 crore and each elephant statue costs Rs. 70 lakh. In fact Mayawati spent Rs. 2500 crores on five such major parks across Uttar Pradesh. But the Noida memorial to Ambedkar has had a chequered history. In September 2009, a Supreme Court committee recommended putting a halt on the construction as the project did not have an environment clearance. In December 2010, the Supreme Court gave a green signal to Ambedkar Park saying the park is

[72]For detail see news under headline "71/80: BJP posts its best-ever show in Uttar Pradesh" on May 17, 2014, http://timesofindia.indiatimes.com/home/lok-sabha-elections-2014/news/71/80-BJP-posts-its-best-ever-show-in-UP/articleshow/35229428.cms, accessed May 17, 2014.

[73]See in detail in news under the headline "Mayawati inaugurates Rs. 650 crore park in Noida" on October 14, 2011 at 05:45 pm, http://ibnlive.in.com/news/mayawati-inaugurates-rs-650-crore-park-in-noida/193078-37-64.html?utm_source=ref_article, accessed October 14, 2011.

Image 4.12 Politics on public exchequer. *Source* Aaj Tak, November 14, 2011 (Mayawati with Satish Mishra)

not on forest land. Environmentalists had alleged the project led to ravaging of Noida's green belt and the loss of 6000 trees. To make matters worse the concrete park has come up right next to Okhla Bird Sanctuary that caused the activity."[74]

The park drew the attention of people, politicians and media since its inauguration and the political debate was shaped around these statues in the media. Media reported on the parks in view of the cost, structure and systematic installation of statues. The issue was also discussed in the media as *inauguration of Dalit inspirational park.*[75] The political debate was discussed in several ways including a discussion around the sense of "social space"[76] and "cultural space"[77] during the election campaign.

---

[74]See in details in news under the headline "'Mayawati inaugurates Rs. 650 crore park in Noida" on October 14, 2011 at 05:45 pm, http://ibnlive.in.com/news/mayawati-inaugurates-rs-650-crore-park-in-noida/193078-37-64.html?utm_source=ref_article, accessed October 14, 2011.

[75]See for details in news under the headline "दलति प्रेरणा स्थल पार्क हुआ उद्घाटन" on November 15, 2011 on Zee News.

[76]For details see Lefebvre (1991). *The Production of Space* (Donald Nicholson-Smith Trans.). Oxford: Blackwell. Lefebvre argues that "Like any reality, social space is related methodologically and theoretically to three general concepts: form, structure, function. The form of social space—i.e. the centre-periphery relationship-has recently come to occupy a place in our thinking about forms. As for the urban form—i.e. Assembly, encounter and simultaneity—it has been shown to belong among the classic forms, in company with centrality, difference, recurrence, reciprocity, and so on" (ibid.: 147). "Social space implies a great diversity of knowledge. What then is its real status? And what is the nature of its relationship to production?" (ibid.: 73). "The form of social space is encounter, Assembly, simultaneity. But what assembles, or what is assembled? The answer is: everything that there is *in space*, everything that is produced either by nature or by society, either through their cooperation or through their conflicts. Everything: living beings, things, objects, works, signs and symbols. Natural space juxtaposes and thus disperses: It particularizes. By contrast, social space implies actual or potential Assembly at a single point, or around that point...... Social space per se is at once work and product-a materialization of "social being." In specific sets of circumstances, however, it may take on fetishized and autonomous characteristics of things" (ibid.: 101–102).

[77]Cultural space is relational space wherein space supposed to represents through a *thing* which is shaping and reshaping the space in myriad ways. The Dalit inspiration park of Lucknow and Noida is somehow representing a period of struggle in contentious ways.

The mediated political discourse around the statues was discussed first, when the English media portrayed the issue in an economic context, while the Hindi media portrayed the same discourse using the caste framework. However, the opposition parties were widely reported in the media criticizing Mayawati for such constructions while BJP spokesperson Ramanath Kavid demanded the removal of statues and said that *statues of Elephant removed from Uttar Pradesh*[78] and asked the ECI to look into the matter in view of election code of conduct. The political discourse around these statues involved the Election Commission of India (ECI) as well. The opposition parties approached the ECI and complained about the similarity between the statues of Elephants and the BSP election symbol. Thereafter, ECI in its judgement passed an order that *statues of Mayawati and the Elephant will be covered*[79] and *until election statues will be covered*[80] with immediate effect in the state till elections were over. The political discourse became so complex when BSP criticized the Chief Election Commissioner's (CEC) decision to cover up the statues and said, "EC order one-sided,"[81] while in his response CEC said, "Why should Mayawati object, when BJP, Congress accepted our earlier decisions?"[82] However, Mayawati termed the judgement passed by ECI as being "arbitrary" and "one-sided." She clarified about her statue and said that it was a part of Kanshi Ram's wish and he "willed that her statue will be installed along with that of his." In her response to the poll symbol, she argued that the installed Elephant in the parks are in *swagat mudra* (welcoming posture) and the media quoted her as saying "They had not been installed by her government in parks and memorials on account of the elephant being the poll symbol of the BSP"[83] but due to the Elephant being a part of the cultural identity and symbol of power in India.

Meanwhile, the Uttar Pradesh administration started to cover up the statues in the parks and the headlines read—"Pink cover for Mayawati, elephant statues to cost Rs. 1 crore."[84] The selection of pink colour was natural because the colour is

---

[78]See for details in news under the headline "यूपी में हटाई जाएं हाथी की मूरतियां" on January 1, 2012 on Aaj Tak.

[79]See for details in news under the headline "यूपी में ढंकी जाएंगी मायावती, हाथियों की मूरतियां" on January 1, 2012 on Aaj Tak.

[80]See for details in news under the headline "चुनाव तक परदे में रहेंगी माया व हाथी की मूरतियां" on January 8, 2012 on Rashtriya Sahara.

[81]See in detail in news under headline 'EC order one-sided: Mayawati' on 15.1.2012 on The Hindu, http://www.thehindu.com/news/national/ec-order-onesided-says-mayawati/article2803608.ece, accessed January 15, 2012.

[82]See in detail in news under the headline "Why should Mayawati object, when BJP, Congress accepted our earlier decisions?" on January 18, 2012 on The Hindu, http://www.thehindu.com/news/national/why-should-mayawati-object-when-bjp-congress-accepted-our-earlier-decisions/article2806282.ece, accessed January 18, 2012.

[83]See in detail in news under the headline "EC order one-sided: Mayawati" on January 15, 2012 on The Hindu, http://www.thehindu.com/news/national/ec-order-onesided-says-mayawati/article2803608.ece, accessed January 15, 2012.

[84]See in detail in news under the headline "Pink cover for Mayawati, elephant statues to cost Rs. 1 crore" on CNN IBN, January 10, 2012 at 04:58 PM IST, Read more

Image 4.13 Politics of election symbol. *Source* Aaj Tak, January 10, 2012

not associated so far with any political party in the state (Image 4.13). The officials were busy to meet the deadline fixed by ECI to cover up all the statues by January 11, 2012. There were more than 200 statues of Mayawati and the Elephant out of which, the maximum number were installed in Lucknow and Noida. The estimated cost of the cover up project was one crore (10 Million) and around 1.5 km of pink polythene has been used for covering the statues.

The decision to cover up the statues was welcomed by the opposition parties in the media while the Congress Party cautiously welcomed the decision and believed that the cover up too would give political benefits to BSP. As Congress spokesperson Rashid Alvi said that *covering the statues will give benefits to BSP*.[85] However, BSP out rightly rejected the decision of the ECI and a section of politicians also felt that the covering of the statues provided Mayawati with an issue to discuss in the polls. However, Mayawati brought forward Satish Mishra to counter the cover up politics and Satish Mishra wrote a letter to ECI where he mentioned the decision as being one-sided and a violation of Article 14 of the Constitution, which provides the Right to Equality. He further said, "Dalits are feeling cheated as the decision was taken on the memoranda of SP, Congress and BJP. The statues installed in Dalit memorials are in welcome position with upside trunk and are not similar to BSP's symbol. Elephant statutes are installed in temples across India and government buildings. The EC should also have objected to SP's symbol cycle, BJP's lotus and RLD's hand pump. The EC should have also taken note of statues of former presidents and leaders of other parties. The statutes of Mayawati, who is '*Dalit ki Beti*' (daughter of Dalits) and has been four time Chief Minister, will not affect voters in any way."[86]

(Footnote 84 continued)

at: http://ibnlive.in.com/news/up-pink-cover-for-mayawati-elephant-statues-to-cost-rs-1-crore/219538-37-64.html?utm_source=ref_article, accessed January 10, 2012.

[85]See for detail in news under the headline "मूरतियां ढंकने से बीएसपी को ही फायदा" on January 15, 2012 on Aaj Tak.

[86]See in detail in news under the headline "Statue cover up gives Mayawati a poll issue" on January 13, 2012 on TOI, http://timesofindia.indiatimes.com/india/statue-cover-up-gives-mayawati-a-poll-issue/articleshow/11470411.cms, accessed January 13, 2012.

Image 4.14 Politics of idealism. *Source* TOI, January 23, 2012, p. 13 (dance of democracy)

In (Image 4.14), a politician's wife is saying to her husband, "EC will cover you up for resembling the BSP poll symbol" when her husband is paying respect to the image of Mayawati. In the cartoon, the posture resembles the BSP's election symbol Elephant. The political discourse around the issue was constructed in such a way that mere resemblance to a Party symbol pushed the EC to cover up the statues. The mediated politics portrayed the entire discourse in a rhetorical and satirical manner when Mayawati complained about other political parties about that how they are using their poll symbol openly to ECI and asked to take necessary action against them. However, in his reply ECI told her (Mayawati) that there are many things beyond his jurisdiction that cannot be controlled. Hence, pink colours of polythene were used to cover up the statues of Mayawati, was depicted in the media (Image 4.15). Mulayam Singh Yadav further aggravated the issue by stating in the media that his government would use the space near the statues for a hospital and school while the BJP leadership was keen to replace them with statues of freedom fighters (Image 4.16). However, since the SP came into power, Akhilesh Yadav told the media persons that *they shall not remove statues of Elephant and Mayawati's*[87] and urged the enquiry committees to look into the cases of corruption and mismanagement of public funds instead.

Once the election was over, the ECI passed an order to unveil all the statues with immediate effect. People addressed these statues in various forms from seeing them as grandeur works of arts to showing displeasure over them. Some people were

[87]See in detail in news under the headline "नही हटाएंगे मायावती की मूरतियां" on February 5, 2012 on Aaj Tak.

Image 4.15 Politics of representation. *Source* TOI, January 17, 2012, p. 8

Image 4.16 Politics of landscapes. *Source* TOI, March 2, 2012, p. 12

amazed that the unused land around river Gomti at Lucknow was allocated for the park, while others felt pride at being Dalits.[88]

In the CSDS survey, nearly half the respondents were of the view that the Election Commission did the right thing by ordering the covering of the statues of Mayawati and the Elephant during the elections. Only 16% said it was a wrong decision. Even among the Dalits, most felt the EC did no wrong. Majority of the

[88]Tiwar and Pandey (2013), Tranquebar Press. pp. 144–45.

Table 4.3 Citizens' opinion on statues built by BSP

Statements	All		Dalits	
	Agree	Disagree	Agree	Disagree
EC was right in covering statues during election	48	16	37	23
BSP government has wasted money on building statues and parks	51	25	45	28

Source CSDS, 2012, (EPW, April 7, 2012. Vol. XLVII NO 14)

respondents (51%) were also of the view that the BSP government had wasted money on building statues and parks (Table 4.3).[89]

However, the public opinion in most cases were that "making statues of a leading person and spending ten millions on them is culturally outrageous."[90]

The politics around the statues was a contentious issue and both the BSP and opposition parties had their own political stand over it. The opposition considered them as misusing public funds while the BSP considered these statues as an acknowledgement of the social struggle and saw the parks/memorials as being a part of the cultural capital to recognize the value of these leaders. The issue shaped the heartland in the political context and symbolized Dalits politics in terms of the cultural space as well.

4.9 Conclusion

During election campaigns media unveiled the "real" and "imagined" spatial politics. Media is a medium to help us to understand the spatial politics at the herd of the region. The dimensions of regional politics changed over time and both media and politics shaped the regional politics as well. As Stahlberg points out, "The expansion of media and simultaneously regionalization of media highlights the regional leader and regional politics in India."[91] The heartland was the product of political consequences of regional politics when regional politics overshadowed the national politics or regional politics and drew the map of national politics. A number of issues surfaced during the elections in the heartland and each issue had its own political consequences and domain of interests.

The announcement of a seven-phase election by the ECI provided political ammunition to the opposition parties to attack BSP. However, the BJP overreacted on the issue while other political parties welcomed the move to conduct free and fair elections in the state. The announcement of the election schedule to some extent

---

[89]Kumar et al. (2012).

[90]Interview with Adarsh Tiwari. He is lawyer by profession and political activist since college days and belongs from Ayodhya (Uttar Pradesh) and living in Delhi last 5 years. He closely watched the Uttar Pradesh election and agreed over the role of media during the Uttar Pradesh election.

[91]Stahlberg (2013, p. 63).

was considered the beginning of the election process and politicians were supposed to hold political rallies and meet to attract crowds and thus, the Helicopter and Bollywood celebrity became political tools to pull people towards the political venue. The announcement of the seven-phase Assembly elections reflects the political sensitivity of the ECI and ECI worked with the impression that the state is the heartland of Indian politics.

During the election campaign, political parties promised to give laptops and tablets (SP and BJP) to the youth voters who were voting for the first time. To some extent, the Uttar Pradesh elections were the turning point when media gadgets were given preference over other freebies. For example, in Tamil Nadu, the announcement of freebies in their manifesto was considered a common practice. However, during the Uttar Pradesh elections, some other issues resurfaced in the media such as the Babri-Ayodhya conflicts, unemployment, old age, women compensation, power project were discussed by politicians across party lines. However, media reported the political statements of senior politicians where they talked on development, corruption, caste, reservation and the backwardness issue. The political manifestoes were shaped to reflect the imagination of the heartland wherein political parties made promises to send the message of a broader political imagination having an inclusive approach.

In every exit poll discussion held in the media, SP was shown as the largest political party and, the Congress Party was thought to do better. Meanwhile, SP won the elections way beyond the exit poll prediction. During the media discussions, the relevance of exit polls was also discussed. For example, during the General Election in 2014, none of the exit polls predicted that BJP will get its own majority in the Parliament while every exit poll showed that BJP might become the largest political party in the parliament. Thus, exit polls have become an integral part of discussion of elections in the media.

The politics of the creation of four new states being carved out of Uttar Pradesh, suddenly changed the political discourse in the state and provided an opportunity to the opposition parties to criticize the BSP government. The politics around the issue was analyzed as being a master stroke of Mayawati and an ambitious political plan. The creation of four new states was based on the regional aspiration and was aimed at raising Mayawati's political stature in national politics. The political discussion in Hindi and English media were very different. The Hindi media acknowledged the politics of the proposed four new states, while the English media described the decision as a political move and emphasizing the timing of the announcement. Meanwhile Hindi media attacked BSP and saw this as hiding the failures of BSP government and shifting the attention of the people prior the elections.

The political debate around the statues of Mayawati and the Elephant made the Uttar Pradesh elections distinguishable from other State Assembly elections and the issue was interpreted as an extension of Dalit politics and the mismanagement of public funds in the media as well. However, BSP acknowledged these statues as being the space for a "Dalit Inspirational Park" and a part of the cultural capital to represent the social struggle of the society. These statues were seen as giving a sense of place to the Dalit people. The English media interpreted the issue along with the fiscal policy and highlighted the insensitive approach of the BSP government too while Hindi media interpreted the statues as being an extension of social identity and matter of social pride in the state. The statues generated a sense of association amongst a community that shaped the political and social context of Uttar Pradesh as heartland in Indian politics.

Media shaped these political discourses in contentious manner with political parties made their public stands clearly in contrast to one another. During the elections, the political discourse can be interpreted as a struggle between national and regional political parties and the ways in which regional political parties overshadowed the national political parties over time. The interpretation of the heartland was mapped according to the ways in which the institutions (media, political party and ECI) addressed the issues. All these institutions mapped the heartland as being a strategic, political state that would have an impact on the impending parliamentary election of 2014.

The next chapter questioned the functioning style of mediascapes with respect to issue "reservation for minority." It was one of the eight images (issues) identified by mediascapes and mediascapes discussed the image as reservation for Muslims instead of "reservation for minority," what was in actual. However, there are six communities come under minority status (Muslim, Christian, Sikh, Jain, Buddhism and Parsi) in India and reservation policy was supposed to announced for all communities, not only for Muslims, but media represented the issue into communal colour.

References

Boyd-Barrett et al (eds) (2012) Approaches to media: A reader. Rawat Publications, Jaipur

Crampton J (2010) Mapping: a critical introduction to cartography and GIS. Wiley-Blackwell, Oxford

Habermas J (1989[1962]) The structural transformation of the public sphere: an inquiry into a category of bourgeois society. Polity Press, Cambridge

Kumar S et al (2012) Sixteen assembly elections in Uttar Pradesh, EPW, April 7, XLVII(14): 80–86

Lefebvre H (1991) (Donald Nicholson-Smith Trans.). The Production of Spaces. Blackwell, Oxford

Lundby K (ed) (2009) Digital Storytelling, mediatized stories: self-representations in new media. Peter Lung, New York

Six C (2010) Spectacular politics: performative nation-building and religion in modern India. Manohar Publications, New Delhi, p 121

Stahlberg P (2013) Writing society through media: ethnography of a hindi daily. Rawat Publications, Jaipur

Tiwari M, Pandey R (2013) BattleGround UP: politics in the land of ram. Tranquebar, Chennai

Chapter 5
Locating the Communal Vote Bank Politics in the Mediated Heartland of Uttar Pradesh

Abstract Vote bank politics is older phenomena in India wherein politicians and political parties are supposed to announce political soup during election time to lure voters, to expand political reach amongst voters as well. Congress Party announced 4.5% reservation for minority prior to Uttar Pradesh election to lure voters and the news would have been discussed by media in a communal colour and interpreted as reservation for Muslims. Hence, politicians are often known for such activities but here media was responsible to map the issue into communal shadow.

Keywords Vote bank politics · Minority · Muslims · Reservation · Communal

This chapter will discuss the vote bank politics[1] in mediated contexts in both formal and informal texts, situations, background and representations. The vote bank politics has been considered a hot topic of discussion amongst voters and politicians particularly during election time. Moreover, media joined discussion as a third party, apart from people and politicians in term of news coverage and content as well and shaped the discussion of vote bank politics in the country.

In every election, political parties go into the election with some vote bank issues, hoping to materialize them. Every political party tries to unfold their vote bank policies in shape of an election manifesto to address the issues concerned. Except the BSP, every political party had prepared a manifesto during the Uttar

---

[1]In Indian conditions, the term "vote bank" is quite popular and during elections every political party uses the term to suit their political understanding. Muslims are generally considered as a vote bank in a broader sense while some other castes, classes, religions, gender, etc. are also included in the lists referred to for political interpretation. "The term *vote-bank* was first used by noted Indian sociologist, M.N. Srinivas (who also coined the terms Sanskritisation and *dominant caste*); in his 1955 paper entitled *The Social System of a Mysore Village*. He used it in the context of political influence exerted by a patron over a client. Later, the expression was used by F.G. Bailey, a professor of Anthropology at the University of California, San Diego, in his 1959 book *Politics and Social Change*, to refer to the electoral influence of the caste leader. This is the usage that has since become popular. Though the term originally referred to voting along caste lines, it was soon expanded to describe vote banks based on other community characteristics, such as religion and language" [http://en.wikipedia.org/wiki/Votebank, accessed February 23, 2013].

S. Moinuddin, *Mediascape and The State*, Springer Geography,
DOI 10.1007/978-3-319-51932-6_5

Pradesh elections and wished to influence the voters as well. The Congress Party announced a 4.5% reservation for minorities under the OBCs (Other Backward Class) quota to influence minorities across the state. The issue was discussed in the media extensively and many political parties including minority criticized the Congress and read this move as part of their divisive politics. The Indian National Congress (INC) led the UPA II government, which issued the notification of 4.5% reservation for minorities under OBCs quota to fulfil the promises made during the 2009 Parliamentary elections.

A Muslim delegation met Rahul Gandhi and reminded him about the promise he had made during the 2009 Parliament elections—to give reservation to minorities in Central Government jobs. Meanwhile, the delegation assured Muslim support to the Congress Party in the election and in response to the expectations of the delegation members, Rahul Gandhi assured them, "What the government has to do, it will do. But you please go to people and talk about UPA's good work."[2]

This chapter discusses the issue of 4.5% reservation for minorities under OBCs quota as well as how the media swayed the context and perspective of the issue and took it away from the real content. The way in which media represented the issue and emphasized the communal perspective in the issue has also been discussed. The issue was on "Reservation for Minorities" but in fact was discussed as Reservation for Muslims in the media during the election campaign.

5.1 Mapping-Mediated Vote Bank Politics

Common perception of the people seems to be that "except BJP, every political party hopes to gain the Muslim vote during elections and therefore political parties were expected to announce some lucrative political plan for the community."[3] The Congress had taken the advantage and announced 4.5% reservation to minorities under the OBCs quota of 27% in Central Government jobs (Image 5.1). The issue became popular in the media being discussed in the context of minority reservation policy and as part of larger vote bank politics of political parties.

Before elections every political party shapes their political strategy and therefore "Cong to focus on Muslim areas ahead of Uttar Pradesh polls"[4] was unveiled by Sriprakash Jaiswal, the Chief of the Uttar Pradesh Assembly election campaign

[2]See details in news under the headline 'Centre bids to push Muslim quota ahead of polls in Uttar Pradesh,' on December 6, 2011 on TOI, [http://timesofindia.indiatimes.com/india/centre-bids-to-push-muslim-quota-ahead-of-polls-in-uttar-pradesh/articleshow/10998647.cms, accessed December 2011].

[3]It was the part of Focus Group Discussion (FGD) when members felt so that Muslims are ever favorite to the political parties during election.

[4]See details in news under headline 'Cong to focus on Muslim areas ahead of Uttar Pradesh polls,' [http://timesofindia.indiatimes.com/india/Cong-to-focus-on-Muslim-areas-ahead-of-UP-polls/articleshow/10467842.cms? accessed October 24, 2011].

Image 5.1 Politics of Muslims appeasement. *Source* Aaj Tak, December 30, 2011 (Rahul Gandhi)

committee, who said "The campaign will look to carry the message of development, harmony and inclusive growth to maximum people. It has been drawn up with that objective in mind." He categorically focused on Muslim-dominated districts and chose seven districts which were considered the hub of Muslim politics to launch Rahul Gandhi. He explicitly emphasized the importance of Muslims in the state and a blue print was prepared for Rahul Gandhi regarding his campaign schedule. Rahul Gandhi was supposed to start the poll campaign from Lucknow but had not decided the venue and the route that he was supposed to follow to attract the voters. The tentative for Rahul Gandhi campaign trail was from Lucknow to Barabanki; Gonda to Deoria; Kanpur to Ramabai Nagar to Jhansi; Faizabad to Sultanpur and Jaunpur to Mirzapur and Sonbhadra; Saharanpur to Muzaffarnagar; Meerut to Mathura; Agra to Bijnore; JP Nagar to Shahjahanpur and Sitapur to Unnao.[5] The route plan was drawn in order to impress the Muslims who constitute 18.49% of Uttar Pradesh's total population.[6] Zee news portrayed (Image 5.2) to represent the issue of reservation for minorities at best[7]—an unknown boy becomes the symbol of Muslims who, along with others, wears a skull cap—a symbol of being a Muslim.[8] Media often used certain signs and symbolic features to indicate Muslims.

Prior to the election, the leadership of the Congress Party was busy giving final touches to the image of Rahul Gandhi including *where* and *how* Rahul Gandhi would start campaigning launched in order to make a political impact in the state. However, the Congress Party chose to launch his campaign in Muslim-dominated

[5]See details in news under headline 'Cong to focus on Muslim areas ahead of Uttar Pradesh polls' [http://timesofindia.indiatimes.com/india/Cong-to-focus-on-Muslim-areas-ahead-of-UP-polls/articleshow/10467842.cms? accessed October 24, 2011].

[6]See, Census of India 2011 (Uttar Pradesh).

[7]For details see the issues on the Uttar Pradesh Assembly Elections 2012. Out of eight issues, the issue of reservation for minorities was one of them. Also see [http://zeenews.india.com/state-elections-2012/up/issues.html/7, accessed November 13, 2011].

[8]The meaning of sign and symbols can understood through semiotic analysis or the systematic study of signs. "Semiotic starts with the assumption that all communication occurs by the exchange of signs, each of which at least one *signifier* (a mental concept) and at least one *signifier* (a physical aspect)" (Bertrand and Hughes 2005).

Image 5.2 Politics of identity. *Source* Zee news (Appendix A: 235), Image of unknown boy performing Namaz

areas (districts) where the Congress Party perhaps thought that they had a chance to get Muslims votes.

In (Image 5.3) cartoonist Ajit Ninan tries to portray an image of Rahul Gandhi in various dimensions to reflect a multifaceted image of Rahul Gandhi to attract three different sections of voters as "a 3-face strategy for three audiences-upper castes, Dalits and minorities."

The plan unveiled the extent to which the Congress party was willing to acknowledge the Muslim voters and was trying to address their issues. With the approaching election dates, Congress mooted the idea of reservation for backward Muslims and in the All India Congress Committee (AICC) meeting, where Rahul Gandhi unveiled the reservation policy and the media highlighted him as saying, *Reservation for Minorities in central government jobs very soon*[9] (Image 5.4).

However, the Congress supported the reservation plan and argued that in many constituencies the Muslim population was around 18% and could influence the voting trends in the state. The move was of course a political plan to isolate the influence of Samajwadi Party and the Bahujan Samaj Party to make a dent in the Muslim vote bank in the state. Subsequently, the media covered Mayawati addressing the people at a function at Lucknow, saying "Muslims must be given separate quota,"[10] despite contradicting the Congress Party over the issue and she further added "Muslims must be given quota on the basis of their population, for which the proportion of OBCS reservation should be increased and a national policy should be formed to ensure that the benefits of quota reach Muslims across

[9]See for detail in news under headline 'केंद्रीय सेवाओं में मुसलमानों को आरक्षण जल्द' on Zee news on October 31, 2011.

[10]See in detail news under headline 'Muslims must be given separate quota: Mayawati' on December 19, 2011 on TOI, [http://timesofindia.indiatimes.com/city/lucknow/Muslims-must-be-given-separate-quota-Mayawati/articleshow/11161851.cms? accessed December 19, 2011].

Image 5.3 Politics of interpretation. *Source* TOI, January 24, 2012. p. 13 (Dance of Democracy)

Image 5.4 Politics of minorities. *Source* Aaj Tak, October 15, 2011 (Unknown images of muslims)

the country. BSP will support any Constitutional amendment required in this regard."[11] Following an election rally at Barabanki, media selectively represents Mulayam's speech as evident from headlines, *Mulayam played Muslim Reservation*

[11]See in detail news under headline 'Muslims must be given separate quota': Mayawati,' [http://timesofindia.indiatimes.com/city/lucknow/Muslims-must-be-given-separate-quota-Mayawati/articleshow/11161851.cms? accessed December 19, 2011].

Card.[12] Azam Khan had promised a reservation policy beyond the imagination of Muslims and included giving free irrigation water, free education, employment or unemployment benefits and improve the supply of electricity and provide special opportunities to women when back in power. Interesting to note how headlines were worded by media irrespective of political affiliations of the politicians. Across party lines any politician who spoke of developing and empowering minority communities, headlines were worded as 'minority love' or wooing Muslims and reservations particularly for the Muslims. The following narratives prove such selective representations that were made by the media. For example, "Digvijay slams Mulayam for so-called minority love"[13] and asked about his "*minority love*" and his proposed 18% reservation to Muslims said "I think Mulayam must realise that giving 18% reservation would require a constitutional amendment and no less. Can a regional party with only 21 MPs deliver what he promises"?[14] Now, "RLD woos Muslims with 8% quota"[15] and Rashtrya Lok Dal (RLD) Chief Ajit Singh said "Almost 70% Muslims in Uttar Pradesh fall into the category of backward groups. We are waiting for the caste-based census to decide the actual number. But the established formula as per the 'Mandal Commission'[16] is to give reservation to half of the backward population, which for Uttar Pradesh should be around 8%."[17] Salman Khurshid in a election rally at Farukhabad where his wife Louis Khurshid was contesting the election from, announced "Congress would provide nine per cent reservation for backward Muslims within the existing OBCs quota in Uttar Pradesh in the event of its forming the government in the state."[18] The BJP criticized the Congress Party for making promises of 9% reservation, if voted to power.

---

[12]See for detail in news under headline 'अब मुलायम ने फेंका मुस्लिम आरक्षण का पासा' on Aaj Tak on January 8, 2012.

[13]See in details in news under headline 'Digvijay slams Mulayam for so-called minority love,' [http://timesofindia.indiatimes.com/city/lucknow/UP-elections-Digvijay-slams-Mulayam-for-so-called-minority-love/articleshow/11574254.cms? accessed January 21, 2012].

[14]See in details in news under headline 'Digvijay slams Mulayam for so-called minority love,' [http://timesofindia.indiatimes.com/city/lucknow/UP-elections-Digvijay-slams-Mulayam-for-so-called-minority-love/articleshow/11574254.cms? accessed January 21, 2012].

[15]See in detail in news under headline 'Now, RLD woos Muslims with 8% quota,' [http://timesofindia.indiatimes.com/home/specials/assembly-elections-2012/uttar-pradesh/Now-RLD-woos-Muslims-with-8-quota/articleshow/11346203.cms? accessed January 3, 2012].

[16]A commission was established under the chairmanship of B.P Mandal to address the socio-economic conditions of OBCs. The commission recommended 27% reservation for OBCs in education and government jobs to uplift the socio-economic conditions of OBCs. However, V.P Singh implemented the recommendation of the commission and it has been implemented since 1993 in the country.

[17]See in detail in news under headline 'Now, RLD woos Muslims with 8% quota' on January 3, 2012 on TOI, [http://timesofindia.indiatimes.com/home/specials/assembly-elections-2012/uttar-pradesh/Now-RLD-woos-Muslims-with-8-quota/articleshow/11346203.cms? accessed January 3, 2012].

[18]See details in news under headline 'Uttar Pradesh: Congress promises 9 pc sub-quota for Muslims' on January 9, 2012 on cnn-ibnlive, [http://ibnlive.in.com/news/up-congress-promises-9-pc-subquota-for-muslims/219283-3.html?utm_source=ref_article, accessed January 9, 2012].

BJP leader Mukhtar Abbas Naqvi said "BJP leaders met EC today. We have raised the issue about the way the Congress is trying to divide Uttar Pradesh on the base of religion and caste to hide corruption. Whatever Salman Khurshid has said on 9% Muslim sub quota is unconstitutional. This is a violation of moral code of conduct." Naqvi further added "This way of making promises in election rally is unconstitutional. This is an attempt to divide the society on lines of religion and caste." He further said "The EC has given assurance that they will take appropriate steps in complains against the Congress party talking of quota for minorities."[19] Meanwhile, Congress had given the examples of the currently successful models in Kerala, Karnataka and Andhra Pradesh over the minority reservation issue. Mr. Salman Khurshid (the then law minister of India) said that if the Congress formed the next government in Uttar Pradesh the same system would be implemented and completely denied that the move amounted to vote bank politics and insisted that it was a matter of social justice. Salman Khurshid further announced that in the 12th Five Year Plan, steps would be taken to give a stipend to all children belonging to the minorities. While addressing a meeting in Lucknow, Mr. Digvijay Singh said that backward Muslims were not getting the benefits of the 27% reservation policy and therefore the Congress-led Central Government would need to adopt such a policy. The Union Steel Minister, Beni Prasad Verma stated "*Quota decision will make SP irrelevant in Uttar Pradesh polls.*"[20] The decision on reservation for minority groups was described as a "gift from the UPA government," and he further added that it would make "the SP will be irrelevant in the Assembly election."[21]

However, Azam Khan dismissed the UPA government's move and instead demanded reservation for the minority community in accordance with their population. However, the SP termed it as an election stunt by the Congress-led UPA government. At the same time, Mayawati demanded a national reservation policy for backward religious minorities. The Peace Party (Uttar Pradesh-based political party) described this as the "biggest betrayal of Muslims since independence."[22] In an interview, Congress spokesperson Mr. Rashid Alvi (Image 5.5) said *it is the least for Muslims*[23] and that there is still a lot more to be done in this area.

---

[19]See in detail in news under headline 'BJP meets EC over Cong's Muslim quota promise' on January 10, 2012 on cnn-ibnlive, [http://ibnlive.in.com/news/bjp-meets-ec-over-congs-muslim-quota-promise/219571-8.html?utm_ source=ref_article, accessed January 10, 2012].

[20]See in detail news under 'Quota decision will make SP irrelevant in Uttar Pradesh polls: Verma on 24.1.2012' on The Hindu, [http://www.thehindu.com/news/national/quota-decision-will-make-sp-irrelevant-in-up-polls-verma/article2742415.ece, accessed January 24, 2012].

[21]See in detail news under 'Quota decision will make SP irrelevant in Uttar Pradesh polls: Verma on 24.1.2012 on TOI, http://www.thehindu.com/news/national/quota-decision-will-make-sp-irrelevant-in-up-polls-verma/article2742415.ece, accessed January 24, 2012].

[22]See in detail news under 'Quota decision will make SP irrelevant in Uttar Pradesh polls: Verma on 24.1.2012 on TOI, http://www.thehindu.com/news/national/quota-decision-will-make-sp-irrelevant-in-up-polls-verma/article2742415.ece, accessed January 24, 2012].

[23]See for detail news under headline 'मुसलमिों के लिए जो हुआ वह नाकाफी: राशदि अलवी' on December 23, 2011 on Aaj Tak.

Image 5.5 Muslims voice. *Source* Aaj Tak, December 23, 2011 (Rashid Alvi)

BJP leader Uma Bharti stated in a press conference at Lucknow, "Congress trying to divide and rule India"[24] (Image 5.6) and she further accused the Congress of trying to bring about "a second partition" for electoral gains, saying "We criticize the attempt of the Congress to divide the Muslims by offering those reservations just before elections." The Congress' eye, she said, was fixed firmly on the Muslim vote bank in Uttar Pradesh.[25]

The Hindu came with the news that "*Muslims groups see 'minorities' quota as a googly*"[26] when the President of Welfare Party Mujtaba Farooq argued "there are studies to show that Muslims currently get a share of about three percentage points in the OBCs quota. With a 4.5% share for all minorities, they will actually stand to lose." Maulana Khalid Rasheed Firangi Mahal, a prominent Shia leader and Imam of Lucknow's Aishbagh Idgah, welcomed the decision but stressed that reservation for Muslims should be according to the population. Zafaryab Jilani, executive member of the All India Muslim Personal Law Board (AIMPLB), said a positive step had been taken, but demanded 9% quota for backward Muslims. Zakat Foundation President Zafar Mahmood said as covered by media that "to reserve 4.5% for minorities without clarifying the Muslim share in this does not improve our situation in anyway. If Muslims, who form the bulk of minorities, had been exclusively given this share, we could have seen it as a beginning. This measly measure is no help at all." Manzoor Alam of The Institute of Objective Studies

[24]See detail in news under headline 'Congress trying to divide and rule India' on January 11, 2012 on NDTV, [http://www.ndtv.com/article/assembly-polls/congress-trying-to-divide-and-rule-india-uma-bharti-165876?curl=1390330414, accessed January 11, 2012].

[25]See detail in news under headline 'Congress trying to divide and rule India' on January 11, 2012 on NDTV, [http://www.ndtv.com/article/assembly-polls/congress-trying-to-divide-and-rule-india-uma-bharti-165876?curl=1390330414, accessed January 11, 2012].

[26]See news under 'Muslim group see 'minorities' quota as a googly' on The Hindu on December 23, 2011. [http://www.thehindu.com/news/national/muslim-groups-see-minorities-quota-as-a-googly/article2742416.ece accessed December 23, 2011].

Image 5.6 Politics of resistance. *Source* NDTV 24*7, January 11, 2012, (Uma Bharti)

(IOS) said that though he was deeply disappointed, he welcomed the announcement as a "first step towards justice to Muslims." He said that Muslims had got their hopes up regarding job reservation for the first time since Independence and pointed out "We were hoping and expecting that the government will go at least half way towards implementing the Ranganath Misra Commission report."[27]

The political anguish over the issue was quite evident and some sections of the minorities were dissatisfied and felt that it was merely an extension of appeasement politics and nothing more than that.[28] The way, in which the reservation policy was brought in the public domain by the media, it was clear that it was aimed at luring a section of the population and it had a populist agenda despite the media covering contradictory versions on the issue either by political parties or from within the Muslim community.

Every political party tried to connect with the Muslims during the elections and in order to do so, they announced some percentage of reservation for Muslims. For example, the Congress announced 4.5% under OBCs quota; SP sought an 18% reservation for Muslims, while the BSP sought that a national policy should be prepared to provide reservation for Muslims. In order to get the Muslim support, every political party seemed to have played vote bank politics to woo the Muslims in the state as covered by the media through the headlines and the content of news that appeared in the political campaigns of parties.

Media seemed to portray battle for the Muslim vote as becoming fiercer with the approaching elections in the state. Muslims can influence in around 130 constituencies out of 403 Assembly seats in the state (*The Times of India,* 18.1.2012). Therefore, the Congress Party had given tickets to 61 Muslims for Assembly seats, SP had given 75 Muslims tickets and BSP had given tickets to 84 Muslims to show the Party's solidarity with Muslims.[29] The media, on the other hand, portrayed

---

[27]See news under Muslim group see 'minorities' quota as a googly on The Hindu on December 23, 2011, [http://www.thehindu.com/news/national/muslim-groups-see-minorities-quota-as-a-googly/article2742416.ece, accessed December 23, 2011].

[28]The narratives discussed during Focus Group Discussions (FGD).

[29]See for details news under the headline "All eyes on the Muslim vote in Uttar Pradesh" on TOI on February 12, 2012. p. 12.

Image 5.7 Empowered Muslim women. *Source* TOI, February 12, 2012, p. 12 (Dance of Democracy)

Muslims through their dress and appearance. Muslim women were shown standing in a queue to cast their votes and displaying their voter identity cards (Image 5.7). Muslim women wearing the *burqua* (veil) and men with a beard, wearing a skull cap, represented their distinct cultural identity.

The media often used "identity"[30] to represent a community in political, social, cultural, geographical contexts. This representation could be interpreted as the Muslim women waiting for their turn to cast their vote and being highly participatory in the democratic processes perhaps in the backdrop of the reservations that have been promised to them. Since the announcement of reservation for minorities, the issue came to be associated in media with the Muslim community, irrespective of the content of the issue. On the other hand, Muslim leaders and intellectuals welcomed the decision and reiterated that there was ample scope to adopt a few more corrective measures to improve the socio-economic conditions of the minorities in the country.

In a survey (Table 5.1), CSDS found that "Less than half the respondents had heard of the recent announcement by the central government of giving reservation to minorities within the OBCS quota. However many of those who had heard about the move supported it (48%)."

[30]The "identity" often assumes as sense of ourselves and "what is not, a marking out of the 'other' and one's difference" (Crossley 2005). For details see Crossley (2005).

Table 5.1 Citizens' awareness and support for 4.5% quota for minorities within OBCS quota

Reservation issue	All	Muslims
Those who have heard of 4.5% quota for minorities within OBCS quota	42	51
Support 4.5% quota for minorities within OBCS quota (amongst those who have heard)	48	60

Source CSDS (2012) (EPW, April 7, 2012. Vol. XLVII NO 14)

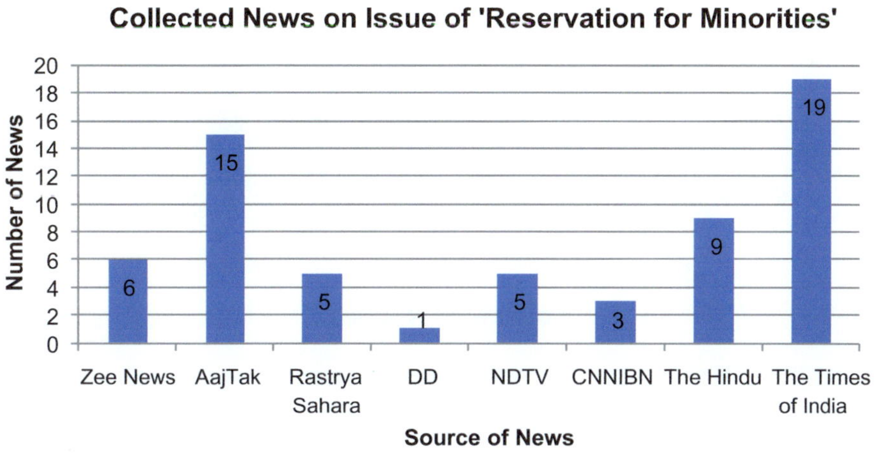

Fig. 5.1 Collected news on issue of 'Reservation for Minorities'. *Source* Primary research survey, 2011–2012

This study had collected 63 news items in the issue of reservation for minorities from different media spaces (Fig. 5.1). Hence, the discussion on the issue varies from one media source to another and most of the sources discussed the issue as reservation for Muslims (Fig. 5.2).

Except *NDTV 24*7* and *Rashtrya Sahara,* all media spaces considered in the study had diluted the issue and discussed it as a Muslim issue rather than one concerning minorities as a whole.

The issue was diluted in both print and visual media. Around 73% news content in *Aaj Tak* and around 83% news content in *Zee News* on reservation for minorities was interpreted as reservation for Muslims. *CNN IBN* discussed the issue as pertaining to Muslims, while *NDTV 24*7* followed the news in its original content and discussed it as an issue regarding the reservation for minorities. DD interpreted the issue as one concerning Muslims in accordance with what the Congress-led UPA government wanted to portray. Furthermore, two noted English newspapers, *The Hindu* and *The Times of India*, interpreted the news as reservation for Muslims 45 and 79% of the time, respectively (Table 5.2). Both print and visual media thus played an instrumental role in shaping and reshaping the issue to present it with communal undertones.

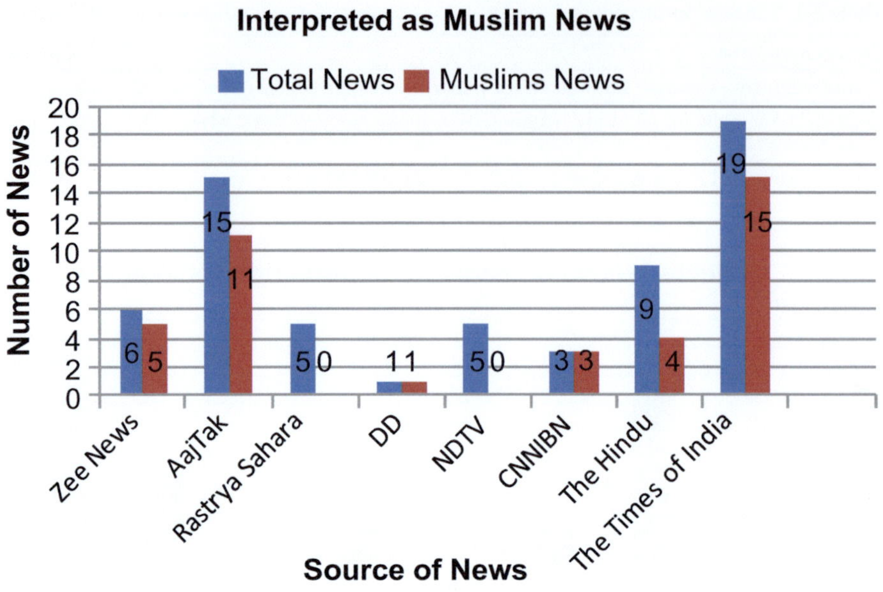

Fig. 5.2 Interpreted as Muslim news. *Source* Primary Research Survey, 2011–2012

Table 5.2 Media interpretation of the news as 'Reservation for Muslims' instead of 'Reservation for Minorities'

S.No.	Source of news	Percentage of news interpreted as reservation for muslims instead of reservation for minorities
1	Zee News	83.3
2	Aaj Tak	73.3
3	Rashtrya Sahara	0
4	DD	100
5	NDTV 24*7	0
6	CNN IBN	100
7	The Hindu	44.4
8	The Times of India	78.9

Source Survey recorded in 2011–2012

5.2 Conclusion

During the elections, almost all political parties extended their political support to the minorities and came out with ways in which to address the minority issue. The political parties were at loggerheads to take credit for the initiative and both regional and national parties had unveiled plans to influence the minorities. The

Congress Party had announced 4.5% reservation for minorities under OBCs quota in Central Government jobs. The SP reiterated this and also promised to provide 18% reservation to Muslims in the state, while the BSP wished for a National Reservation Policy for Muslims. However, the BJP criticized both the Congress led and the SP for their offer. Despite this, the Congress Party leadership reiterated its promise for 9% reservation to the Muslims, if it came to power. Politicians discussed this issue with reference to minorities as well as Muslims.

The media, on the other hand, constructed the issue by giving it a communal colour, as it discussed the issue mostly in light of Muslim reservation. It is commonly believed that politicians spread hatred between communities but in this case, the media became instrumental in changing the aim of the issue and moulding it around Muslims instead of minorities.

The forthcoming chapter discussed and portrayed development and governance issues along with caste and communal politics. This chapter is supposed to looking four issues: Rahul Gandhi and his political campaign; politics of land agitation; politics of NRHM and anti-incumbency politics. Mediascapes discussed these issues as national flag ship programme, development and governance in one hand, while on other hand alleged that state government diluted these programmes and not taken appropriate measure to strengthen the development and governance issues at best in the state.

References

Bertrand I, Hughes P (2005). Media research methods: audience, institutions, texts. Palgrave Macmillan, New York, p. 202

Crossley N (2005). Key concepts in critical social theory. Sage Publications, London, p. 144–47

CSDS (2012) Sixteen assembly elections in uttar pradesh. EPW XLVII(14):80–86

Kumar S et al (2012) Sixteen assembly elections in Uttar Pradesh". EPW 47(14):80–86

Chapter 6
From National to Regional: Development and Governance as the New Avatars of Uttar Pradesh Heartland Politics

Abstract During political campaign mediascapes portrayed development and governance issues along with traditional issues or in other words, it tried to eclipse negative political discussion on caste and communal politics. National leadership alleged the incumbent state government for diluting the flagship programme—are supposed to improve infrastructure and health condition in the state. Regional parties too alleged the national parties for divisive policy that harmed the cultural ethos in the state.

Keywords Development · Governance · Anti-incumbency · Backwardness

When development[1] becomes a strategic issue across political parties and politicians, drawing politics around the development issue can be understood as development politics. Political debates and discussions over development issues were witnessed during the Uttar Pradesh Assembly election campaign and particularly the manner in which the media represented these discussions and party positions through the coverage of news and its contents.

Two key development issues appeared during the Uttar Pradesh election campaign: one pertaining to the land agitation of farmers (Bhatta Parsual) while the other was the NRHM scam (Kushwaha fiasco). The issue of Bhatta Parsual was a land agitation against forced land acquisition by the Uttar Pradesh government while the NRHM scam was a mismanagement of public funds meant to improve the infras-

[1]Development is a broader meaning while in geography it used as purposefully either to indicate the development or to evaluate the development. "Development is used in everyday speech to refer to change. This change is usually viewed in positive terms. However, within geography, development usually has more specific meanings, referring to either national-level processes of economic, political and social change, or the positive change resulting from intentional actions to improve the living conditions of poor or marginal populations. As well as being a process, development can defined as a state of being, usually applied to a country or region and implying high levels of urbanization, complex economic activity and standards of living. Such definitions are, however, not neutral as they reflect particular ideologies which vary across time and space. Geographers have been involved in both reinforcing particular concepts of development and revealing the ways in which they are based on the operation of power".

S. Moinuddin, *Mediascape and The State*, Springer Geography,
DOI 10.1007/978-3-319-51932-6_6

tructure of hospitals and medical conditions in the state. However, both the issues attracted the attention of people and politicians including that of Rahul Gandhi and the media as well. The Bhatta Parsual land agitation drew the attention of Rahul Gandhi and he visited the place to show his party's solidarity with the issue which was at one point considered a potential issue for the forthcoming Assembly elections. However, it soon lost its prominence due to political apathy. Also, the issue was settled by the time the Uttar Pradesh Assembly election notification was announced.

The NRHM issue also drew the attention of politicians and the media. These development issues shaped the anti-incumbency wave against Mayawati, wherein the role of the media was instrumental. These issues became a bone of contention between the national (Congress Party and BJP) and regional political parties (SP, BSP and RLD) during election time. Hence, the media discussed both the issues in political contexts along with the politicians who summed up the whole political discourse and used it to facilitate the Anti-incumbency wave against the BSP.

Rahul Gandhi took charge of the Congress Party to establish the Party in the state where it had been out of power for more than two decades.[2] He emerged as a popular political campaigner and the media followed him very closely and reported his remarks which he made either accusing opposition political parties (especially BSP, SP and BJP) or often raising his sleeves during his speeches, adopting an angry posture. Rahul Gandhi raised the issue of backwardness during the election in various ways and the media shaped the election issue as "Rahul Gandhi: A Tireless Campaigner", who tried his best not only for the Congress Party but also for the people of. However, he raised various developmental issues and forced the other political parties to discuss them.

6.1 "Utho, Jago, Badlo": Locating Rahul Gandhi and the Politics of Change

The Uttar Pradesh Assembly election created a precedent as for the first time in the country, the media made eight potential election issues for the election.[3] The issues were politically contentious in nature and were discussed in the media extensively. The role of the media, particularly the national media during Uttar Pradesh elections was criticized for supporting the Congress Party and its leaders, especially Rahul Gandhi for either giving him more coverage or not giving proper coverage to the regional parties during the elections. That the media gave more coverage to Rahul Gandhi during Uttar Pradesh assembly election was of course a business strategy to get more readers and viewership irrespective of caste, class, religion, gender and

---

[2]N.D. Tiwari was the last C.M of the Congress Party in the state when Congress Party was out of power in 1989 and since then Congress never return in the power, though Rahul Gandhi tried his best to reinstall the Congress Party in the state but failed to make any political impression among people.

[3]See Chap. 3 for details.

Image 6.1 Politics of legacy. *Source* Aaj Tak, November 14, 2011

Image 6.2 Politics of road shows. *Source* The Hindu, February 17, 2012 (Rahul Gandhi greets supporters from a bus during a road show in Lucknow on Thursday—*Image* AP)

regional affiliations. Rahul Gandhi was relaunched in Uttar Pradesh politics and decided to take the responsibility of the Uttar Pradesh elections to improve the performance of his Party in the state. The Congress Party has not returned to power since 1989 in Uttar Pradesh. Rahul Gandhi made an intensive election rally and addressed 211 rallies in a span of 42 days, and covered 375 out of 403 Assembly constituencies since he started his first election rally from Phulpur (Image 6.1) in Allahabad on 14 November 2011 and continued till 29 February 2012 in Bareilly while he also conducted 18 'road shows' in cities like Lucknow, Kanpur and Bareilly.[4] During his road shows (Image 6.2), he greeted the common people as depicted by the media. News coverage too reflected the same as the following narratives of news headlines and content of news indicate:

"Congress general secretary Rahul Gandhi staged a "road show," dubbed as "Jan Sampark Yatra" (door-to-door campaign), here on Thursday to drum up support for

[4]Tiwari and Pandey (2013, p. 39).

Image 6.3 Politics for change. *Source* Aaj Tak, February 5, 2012

party candidates contesting the Assembly polls in the state capital. Riding on a bus embossed with the party's poll slogan, "*Utho, Jaago, Badlo*" (arise, awake and change)[5] (Image 6.3) the Amethi MP's road show stretched for about 34 km across localities, some of them very congested, in all the five Assembly segments—Lucknow Cantonment, Lucknow West, Lucknow Central, Lucknow East and Lucknow North-before rounding off the yatra at the VVIP Guest House. Mr. Gandhi waved from a 'VVIP bus' as a cheering crowd of supporters—comprising youth, mostly from the Old City areas—lined the streets."[6]

However, the road show was a four-and-a-half hour long show and the caravan moved into the Muslim dominated areas of the Old city of Lucknow. In an interview, one of the residents of Lucknow, Yameen Hussain, a trader of Nakhas stated, "People want a clean government, and the Congress should be given a chance to form the next government; both Rahul Gandhi and Priyanka Vadra have been working very hard." Though Rahul Gandhi may have a clean image among the people they were still suspicious that after the elections he (Rahul Gandhi) may not return to the state and concentrate on national politics. Hence the people remained open to alternative options as they saw Rahul Gandhi as a prospective PM candidate of the Congress Party in the forthcoming Parliamentary National Elections of 2014.

Since Rahul Gandhi took over the responsibility of the Uttar Pradesh Assembly elections, it generated a fresh energy in the cadre and state leadership in the state and everyone hoped for the best result in the election.

The Congress Party ruled in the state for almost four decades after Independence without any major political break except in the year 1967. After the withdrawal of the Emergency, the Congress Party was out of power for about 3 years. Thereafter, the party regained its lost political ground. Both Indira Gandhi and Rajiv Gandhi

[5]See Chap. 4 for detail.

[6]See for detail under news headline 'Rahul drums up support in Lucknow' on 17.2.2014 on The Hindu, http://www.thehindu.com/news/national/other-states/rahul-drums-up-support-in-lucknow/article2901075.ece, accessed February 17, 2012.

represented the state in the Parliament till the last elections contested by them. Hence the historical and political significance of the State for the Congress was immense.

This was particularly highlighted by the media. Ajit Ninan, in his cartoon shows the Congress Party leadership as preparing an "IRSO-Indira-Rajiv-Sonia orbit," wherein the scion of Indira-Rajiv-Sonia, Rahul Gandhi was to take the front seat with a major political role (Image 6.4). Before the Uttar Pradesh elections, he was made Vice-President in the Congress Party and assumed the role of General Secretary in the All India Congress Committee (AICC). The Congress was portrayed by the media as heavily relying on Rahul Gandhi and was waiting to see his political calibre and political understanding during the elections. He addressed many political rallies and made extensive political tours in the state.

In all his speeches covered in the media, Rahul Gandhi made it clear that Uttar Pradesh had increasingly become an economically backward state and alleged that all the political parties (SP, BSP, and BJP) were to blame for this. He praised the UPA I and II Government's economic packages given to weavers and the Bundelkhand package. He further discussed how both the BSP and SP had economically looted and pushed the state into backwardness. Rahul Gandhi consistently raised the issue of backwardness throughout his election rally. He made backwardness one of the issues with which to attack his political opponents in the election. He tried his best to gain the people's support and belief in him and even stated in one public meeting at Fatehpur that *I am 41-year-old, will not make false promises.*[7] During his political rally, he was shown black flags in some places by the

Image 6.4 Politics of existence. *Source* TOI, February 8, 2012, p. 13 (dance of democracy)

[7]See for detail news under headline '**41 साल का हूं, झूठे वादे नहीं करूंगा**' on 14.2.2012 on Rastriya Sahara.

supporters of SP and Baba Ramdev but he tried to impress people emotionally and acknowledged his age and further said that he shall not make any false promises.

In his debut election rally in Phulpur, Rahul Gandhi the media portrayed him as addressing various issues primarily economic backwardness of Uttar Pradesh and alleged that other political parties policies had been exploitative in nature. He acknowledged that the rest of the Indian states were growing and prospering while Uttar Pradesh was gradually sliding down in every aspect of development. He urged the youth to come forward to change the system. The media often showed him in anguish against corruption and economic backwardness of the state and covered headlines that said '*I am angered by Uttar Pradesh's situation*'[8] and further shown as saying '*India is growing very fast while Uttar Pradesh is not. Here, criminals are in politics and I have been in politics since seven years and in these seven year have myself travelled a lot to the various parts of Uttar Pradesh and the country. You people taught me lot in these seven years.*

6.2 Mapping the Politics of Backwardness

During the election, the issue of backwardness was discussed across the political parties and people was particularly emphasized by the media. The issue became the bone of contention when both opposition political parties (SP, BJP, and Congress Party) and the governing political party (BSP) failed to agree over the issue. All the political parties appeared in media including the Congress Party and others of reinterpreting the issue of backwardness to gain politically over the governing political party. However, the BSP being the governing political party in the state refuted such charges levelled by opposition political parties over the backwardness issue.

Rahul Gandhi was seen in the media reports, from the very beginning attacked both the BSP government and Mulayam Singh Yadav for neglecting the development issue and the media carried headlines of him directly accusing Mayawati and Mulayam of failing to deliver promises made to the people.[9] Media reports highlighted how Rahul Gandhi from day one took an offensive stand against the issue of backwardness and made political efforts to materialise the issue in favor of the Congress Party and the Central Government. In order to impress people, Rahul Gandhi visited a few districts of eastern Uttar Pradesh to gauge the real problems in the region. He went to all possible districts where he could make an impression. He started his Uttar Pradesh mission even before the election, almost a year beforehand, to touch the emotion chords of the people. He visited Bhatta Prasual village after the police atrocities to console the people.

The media extensively covered Rahul Gandhi's visit to Azamgarh and the Varanasi–Mirzapur–Bhadohi belt in eastern Uttar Pradesh in July and October 2011

---

[8]See for detail news under headline 'यूपी के हालात पर गुस्सा आता है' on 14.11.2011 on Aaj Tak.

[9]See for detail in news under headline 'माया-मुलायम ने जनता को ठगा' on 24.11.2011 on Aaj Tak.

where he spent time with the weavers. During his tour, he came to know that high yarn prices were putting pressure on the handloom sector, which was already facing trouble from power loom units and textile mills and he said, "Listening to the bunker (weavers) community in the villages, on both occasions, I was informed that credit is also not easily forthcoming to these villagers."[10] Therefore, he pleaded on behalf of the weavers—"*Rahul seeks Rs 3000 crore sop for weavers*" to the Central Government to release the package announced in the last budget which had not been implemented yet. The decision of releasing such huge funds before elections was politically criticized and seen as politically motivated. During one speech Rahul Gandhi said "Today, poor people are aware. They know what is good or bad for them. The real problem is that government is not reaching you, those in Uttar Pradesh and MP (Madhya Pradesh)" and further "blames Uttar Pradesh, Madhya Pradesh government for Bundelkhand's backwardness."[11] The focus on Bundelkhand was not new for Rahul and he blamed the BSP government for neglecting the region. He further pushed the Central Government to approve Rs 7000 crore packages for development for the Bundelkhand region.

In 'Mission 2012' media coverage Rahul Gandhi was depicted by the media as pro-development and pro-farmer. In Farukhabad on his 4th day of the 5-day tour in the state media coverage highlighted how Rahul Gandhi emphasized on the mis-management of funds—*development fund looted in Uttar Pradesh*[12] and hoe certain political party were *against the farmer*. He further added that the opposition party was stalling Foreign Direct Investment (FDI) on retail sectors and that they (op-position) did not wish for the poor farmers to earn more. Narratives in the media emerged where Rahul Gandhi in his speech said that even a 6-year-old boy was scared and felt that there is no future in the state and that whatever funds were sent by the Central Government were being eaten by an Elephant. He told the people that political leadership were making a fool of the people and were deliberately not emphasizing computer and English education while they themselves knew both computers and English very well. However, he added that a rich person could demand computer and English education, whereas a poor person demanding the same might become a wrong aspiration.[13]

The media not portrayed Rahul Gandhi as pro-poor but highlighted how Rahul Gandhi emphasized the ability of congress as the lone party to deliver, whereas

---

[10]See in detail in news under headline 'Rahul seeks Rs 3000 crore sop for weavers' on 19.11.2011 on TOI, http://timesofindia.indiatimes.com/india/up-poll-vault-rahul-seeks-rs-3000-crore-sop-for-weavers/articleshow/10788943.cms, accessed November 19, 2011.

[11]See in detail in news under headline 'Rahul Gandhi blames Uttar Pradesh, Madhya Pradesh government for Bundelkhand's backwardness' on 9.12.2011 on TOI, http://timesofindia.indiatimes.com/india/Rahul-Gandhi-blames-UP-MP-govts-for-Bundelkhands-backwardness/articleshow/11039631.cms, accessed December 9, 2011.

[12]See for detail in news under headline 'वकिास के पैसे की राज्य में लूटखसोट' on 16.12.2011 on Zee news.

[13]See for detail in news under headline 'वकिास के पैसे की राज्य में लूटखसोट' on 16.12.2011 on Zee news.

other parties failed despite having developmental programmes. In media report Rahul Gandhi was shown as reminding the people about the "India Shining"[14] campaign of the BJP during 2004 and its content while the farmers were being attacked for demanding fair compensation. He indirectly attacked both Mulayam Singh Yadav and Mayawati and said that both had been negligent towards the poor. Rahul Gandhi believed that Congress Party was a party that would be benevolent towards people from all different sections of society. He reiterated this point and pledged to the people that within 5 years he would change the face of Uttar Pradesh. Rahul raised questions on backwardness in all his rallies and sought the people's blessing during the election. In a rally in Rae Braeli both Sonia Gandhi and Priyanka Gandhi Vadhra attacked the present government for pushing the state into darkness while it was the first moment when both of them had come together to campaign in the state (Image 6.5). However, in some places he faced some agitation and was shown black flags.[15]

The unfurling of the national flag is considered a proud moment for all individuals and groups. The tricoloured flag is a symbol and carries feelings of patriotism and identity (nationalism). When someone is seen carrying or waving a black flag, it is considered that he/she is expressing his/her opposition and anguish against someone or something. During the Uttar Pradesh Assembly election campaign, people were seen showing back flags to Rahul Gandhi on different occasions to register their anguish. A few supporters of the SP also waved a black flag to Rahul Gandhi near his helipad when he appeared from helicopter to address an election rally at Phulpur (Allahabad). The supporters were students of Allahabad University and demanded an intervention from Rahul Gandhi to ensure Allahabad University Student Union elections were held as early as possible.[16] However, an unprecedented incident took place when the police, which was accompanying him for security purposes, along with some Congress leaders including the Central Minister, rushed towards the protesters and beat them (SP supporters) for breaching the security resulting in a scuffle. The media acknowledged the whole episode as *Black flag to Rahul*[17] (Image 6.6). The media widely covered the incident showing, the involvement of Central Ministers. Meanwhile, a police case was lodged against Union ministers and Congress Uttar Pradesh state leader Pramod Tiwari. However, the leaders took the matter to the High Court which also gained wide reporttage—*Congress leaders approached*[18] honourable Allahabad High Court for earliest disposal of the case.

---

[14]An election campaign of BJP launched during 2004 Parliamentary election, Pramod Mahajan was the architect was of India Shining programme, though the campaign was flop and BJP loss the election to the Congress led UPA.

[15]See for detail in news under headline 'वकिास के पैसे की राज्य में लूटखसोट' on 16.12.2011 on Zee news.

[16]Tiwari and Rajan (2013, p. 40).

[17]See for detail in news under headline 'राहुल को काला झंडा, सांसदों ने की पिटाई' on 14.11.2011 on Aaj Tak.

[18]See for detail in news under headline 'राहुल गांधी रैली कांड: कांग्रेस नेता पहुंचे हाईकोर्ट' on 18.11.2011 on Aaj Tak.

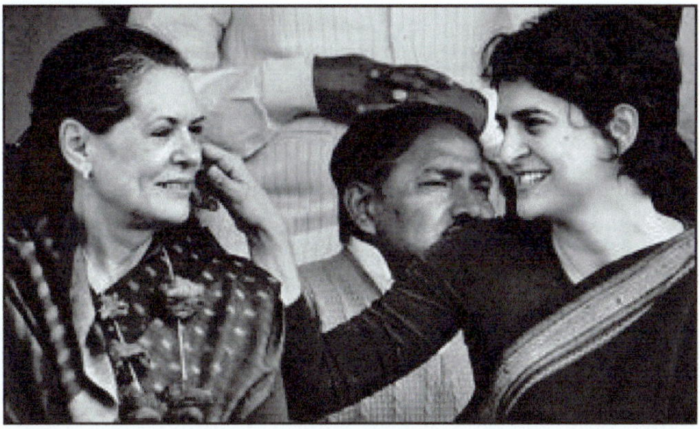

Image 6.5 Politics of gaze. *Source* The Hindu, February 15, 2012 (congress president Sonia Gandhi smiles as her daughter Priyanka Vadra pinches her cheek during an election campaign in Rae Bareli, Uttar Pradesh-both criticized Mayawati)

Image 6.6 Politics of black flag. *Source* Aaj Tak, November 14, 2011

Another incident of black flags being waved took place when a few supporters of the opposition party tried to disrupt the rally and wave black flags and placards[19] (carrying the text—'*we are not beggars*') when Rahul Gandhi was supposed to start his speech in Padrona on 26 November 2011.[20] The black flag chased him across the regions in the state—he was shown black flags in Orai (Jalaun district,

[19]Rahul Gandhi in his political rally at Phulpur (Allahabad) said that the people of Uttar Pradesh are beggars and going other states (Mahrastra, Punjab) for "begging". However, the remarks was politically criticized while Rahul Gandhi explained his motives was economical neither to hurt nor to underestimate the potential of the peoples.

[20]See for details in news under headline 'पडरौना मे राहुल की सभा मे हंगामा' on 26.11.2011 on Aaj Tak.

Bundelkhand) on 18 January, in Duddhi (Sonbhadra district, Poorvanchal) on 22 January, in Kanpur (Awadh) on 20 February, in Bilaspur (Rampur district, Western Uttar Pradesh) on 27 February, in Bareilly (Rohilkhand) on 29 February, in Azamgarh (against Batla encounter case) and many other places including by Baba Ramdev supporters.[21] The supporters of the yoga guru Baba Ramdev on January 19, 2012, disrupted Rahul Gandhi in Banda and thereafter his rally was cancelled when they were shown black flags which caused disturbances at the venue. This incident led to other similar incidents when supporters of Baba Ramdev blackened a poster of Congress president Sonia Gandhi in front of the AICC headquarters in New Delhi.[22] In Aligarh, some supporters of the SP waved black flags at Rahul Gandhi against price rise and corruption on 13 December when he was travelling to Bhim Nagar to address a political rally from Aligarh. Interestingly, in media Rahul Gandhi was the only political leader against whom supporters of opposition political parties waved black flags during the election campaign for political reasons.

The black flags episode during the elections was given due coverage by the media as a form of protest against the congress party and its government at the centre. For example, in an interview Abhishek Yadav, the student union leader of Allahabad University categorically mentioned that if Rahul Gandhi would have taken cognizance of student election issue of Allahabad University, the administration would have conducted Student Union elections on schedule. While contrary to the Allahabad protest, the protests in Aligarh were about the price rise and corruption issues for political mobilization and benefits of the local/regional level of Uttar Pradesh. Every political party knew that Rahul Gandhi, being a high profile leader of Congress Party, could influence the Congress Party's decisions. However, the media portrayed him as an angry man who often raised his voice to speak his mind and criticized the policies of Congress Party to which he belonged, depicting him as a man of the common people and for the common people—A man of the masses.

6.3 Development and Governance: The New Inspirational Dreams of Uttar Pradesh

In the political arena, there was an eagerness to know the capability of Rahul Gandhi and his political skills as well and whether or not he would prove to be successful in bringing the Congress back in the power in Uttar Pradesh after a long period. It was now time for both the Congress Party and Rahul Gandhi to gain the

[21]Tiwari and Pandey (2013, p. 41).

[22]See in detail in news under headline 'Ramdev supporters disrupt Rahul's rally', http://ibnlive.in.com/news/up-polls-ramdev-supporters-disrupt-rahuls-rally/222376-37-21.html?utm_source=ref_article, accessed January 19, 2012.

Image 6.7 Politics of allegation. *Source* Aaj Tak, February 15, 2012

people's confidence back and, therefore Rahul Gandhi touched on almost all issues and showed his plans of the future nature of governance: transparency in governance, inclusive growth and development would be given high priority if he and the congress party were voted to power.

He promised the people of Uttar Pradesh a transparent government if he was voted to power. Rahul Gandhi showed many dreams to the people in his political rallies and promised to prioritize the issues of development, social harmony, as well as issues of cultural and economic concerns to strengthen the system and make the state prosperous. He urged the people to believe him and what he promised along with what he said about other opposition leaders who had been exploiting the state's resources for personal use. These were the main highlights of the media coverage of Rahul Gandhi's political campaign. The following narratives from media reports further substantiate the argument being made here. During his Lucknow rally, he tore down the manifesto of a regional party and said that only a manifesto cannot run the state and rather there was a need for strong determination and transparent government for an inclusive development agenda, and for this, the Congress Party was the only option (Image 6.7). In his 42 days exhaustive election rallies, he covered around 375 constituencies out of 403 in the state and tried his best to put his word across to the people.[23]

In an election rally at Barabanki on 23 November 2011 Rahul Gandhi asked for 5 years and said that *Uttar Pradesh would become developed state in five years*[24] and could end the administration's apathy towards poor farmers in the state. In his speech he accused both BSP and SP for running caste based politics and not focusing on development. He cited the example of Hyderabad and Bangalore and other states where people of Uttar Pradesh generally prefer to go for employment,

[23]Tiwari and Pandey (2013, p. 39).

[24]See for detail in news under headline 'यूपी को पांच साल में बना देंगे अव्वल राज्य' on 23.11.2011 on Zee News.

Image 6.8 Politics of expression. *Source* Aaj Tak, February 6, 2012

and further added that the state must be developed as being parallel to these cities. He promised the people that *Uttar Pradesh can become Maharashtra and Delhi*[25] and further added that the *nation cannot shine without Uttar Pradesh*[26] and alleged that funds were being mismanaged under the MNREGA scheme by the Central Government of India. In an interview to the press in Varanasi on 6 February 2012 Rahul Gandhi expressed his feeling, ambition and his vision of Uttar Pradesh as well as what he had in mind for the state. Rahul Gandhi said that *my mission is to change Uttar Pradesh*[27] and further added that he came to Uttar Pradesh to have a relationship with the people of Uttar Pradesh and not with any political parties and would like to bring about development in the state. Rahul Gandhi addressing a press conference at Varanasi on February 6, 2012 talked on various issues (Image 6.8). He said that while many of the leaders were aspiring for the post of Prime Minister, he did not have such a desire and rather wished to work with the people. He talked about his belief that the ultimate power lay in the people and not in the post. He reiterated that his only aim was to make the development of Uttar Pradesh possible and put forth people's issues and concerns in the parliament. He further added that his aim was to bring Uttar Pradesh on the right track and make a sustainable state and until he achieved his aims he would be around in the cottages of the poor farmers in the state.

In response to the Lokpal question, he said that when he proposed constitutional Lokpal, the BJP out rightly rejected it as it was his offer. In response to black money, though he avoided a direct answer and said every time that someone or political

---

[25]See for in detail in news under headline 'यूपी को दलि्ली, महाराष्ट्र बनाकर दखिाएंगे' on 26.12.2011 on Zee news.

[26]See for detail in news under headline 'यूपी के आगे बढ़े बनिा नहीं चमकेगा देश' on 27.12.2011 on Zee news.

[27]See for detail in news under headline 'यूपी को बदलना ही मेरा मशिन' on 6.2.2012 on Aaj Tak.

Image 6.9 Politics of development. *Source* Aaj Tak, February 5, 2012 [Sam Pitroda (*Left*) and Kapil Sibal (*Right*)]

parties were sending four–five people to disrupt his meeting. Rahul Gandhi further said '*I will remain here whether shows black flags or beat me or hit me with shoes*'.

In a question on Sam Pitroda, Rahul Gandhi was asked why he (Sam Pitroda) kept emphasizing on the fact that his father was a carpenter. Rahul Gandhi in his answer said that caste does matter in Uttar Pradesh politics and that according to Sam Pitroda anyone can bring about change in the system (Image 6.9).

When questioned about Uttar Pradesh's division, Rahul Gandhi completely rejected Mayawati's theory of a partition of the state without any substantial reasons, while there were experts within the country who were aware of the problems and prospects of any region. It was also suggested that Mayawati must first consult these experts before making such a decision. Rahul Gandhi also criticized both regional parties for pushing the state into darkness and showing lack of courage and willingness to bring development in the state. He promised that he would stay in the state and *shall not leave Uttar Pradesh*[28] vowing to change the political direction of Uttar Pradesh. Media carried numerous reports on the lines of change that Rahul Gandhi would bring with particular emphasis in collective responsibility (Image 6.10).

In a survey conducted by Badri Narayan, a social scientist, it was revealed that Rahul Gandhi had the image of an "honest, dedicated and conscientious leader is emerging among peoples of every caste and class."[29] Hence, media played an important role in making successful his image among the people into someone who could be trusted by them and was pro-poor, pro-development and honest in his efforts.

In a CSDS survey it was revealed "that the most popular leader in Uttar Pradesh at present is Mulayam Singh Yadav and by a wide margin. While 33% wanted him to be the next chief minister of the state, 24% wanted Mayawati to continue as chief minister. Rajnath Singh and Rahul Gandhi were the choice of 8 and 7% voters

---

[28]See for detail in news under headline 'यूपी छोड़कर नहीं जाऊंगा' on 13.2.2012 on Rastrya Sahara.
[29]Narayan (2012).

Image 6.10 Politics of face value. *Source* Aaj Tak, November 11, 2011

Table 6.1 Trend in popularity of major political leaders as most preferred CM (2002–12)

Chief Minister Choices	2002	2004	2007	2009	2012
Mulaya Singh Yadav	27	33	28	25	33
Mayawati	19	23	30	26	24
Rajnath Singh	21	7	2	10	8
Rahul Gandhi	NA	1	7	4	7
Akhilesh Yadav	NA	0	1	1	2

Source CSDS (2012)

respectively. Akhilesh Yadav was the choice of 2% of the respondents."[30] However, Rahul Gandhi did not stand anywhere in the race for the post of CM because (Table 6.1) people knew that he had never joined regional politics and rather was a national leader of the Congress Party (Table 6.2).

"Another reason for hope for the Congress is that more people have faith in Rahul Gandhi as a future leader compared to Akhilesh Yadav, Jayant Chaudhury and Varun Gandhi. When asked who among the four would make the best leader in the future, 38% said Rahul Gandhi and 20% opted for Akhilesh Yadav. Even among SP voters, Rahul Gandhi did rather well on this question."[31] People of the state acknowledged the efforts made by Rahul Gandhi but their (people of Uttar Pradesh) preference was to give local leadership a chance though Rahul Gandhi was

[30]Kumar et al. (2012).

[31]Kumar et al. (2012).

Table 6.2 Citizens' choice of future leader

Those Who have faith in...	All	Among Congress voters	Among SP voters	Among BJP voters	Among voters of other party
Rahul Gandhi as future leader	38	64	31	33	37
Akhilesh Yadav as a future leader	20	7	36	12	15
Jayant Choudhary as a future leader	4	8	2	3	4
Varun Gnadhi as a future leader	7	3	3	23	4

Source CSDS (2012)

suitable for the post, he (Rahul Gandhi) was seen as having a preference for national politics over regional politics. Thus, Rahul Gandhi failed to impress the voters to vote for the Congress Party. He had, however, generated a fresh energy among the voters for the Parliamentary elections 2014 wherein people were expecting him to be one of contenders for the post of Prime Minister.

Rahul Gandhi's failure in the state of Uttar Pradesh was interpreted in the media as "may be unlearning" from the past, and strengthening the Congress Party to come back to power (Image 6.11). But many of the people believed that Rahul Gandhi being pushed by the media and perhaps lacked the qualities required as a leader. Many interpreted his earnestly that was portrayed by the media as "signs of immaturity". Thus easily doubts on his ability to lead.[32]

The presence of Rahul Gandhi in the Uttar Pradesh Assembly elections 2012 was interpreted in the media in three ways: as a Prime Ministerial Candidate, Uttar Pradesh Assembly election as a prelude to the General elections 2014 and the rise of developmental and a central concern in Indian politics.

The Uttar Pradesh Assembly elections 2012 were discussed in the media including combinations including the future of the UPA-II government where SP would support the government from outside, the question remained whether BSP would join UPA-II; and whether the government would fall after the Uttar Pradesh Assembly elections and whether there was a possibility of midterm Parliamentary election. In fact, all these discussions shaped the image of "Rahul Gandhi: A Tireless Campaigner"

[32]Interview with Rohit Panchpuri. He is science graduate and political activist. He is vocal on various issues including Rahul Gandhi's. He openly criticized and further said that, Rahul is not a natural politician rather he have been forced to involve.

Image 6.11 Learning the lesson. *Source* The Hindu, March 7, 2012

6.4 Developmental Politics of Land Agitation of Bhatta Parasul

This section is engaged with the issue of development: how the land agitation of farmers' became a curse for SP and how Rahul Gandhi failed to retain people's faith as in the initial phases of the land agitation but gradually lost due to a lack of political vision. The land agitation of Bhatta Prasual had played an instrumental role in shaping the political attitude of the people particularly farmers and poor who were surviving through small land holdings, against SP and "reducing the once 'would-be PM' to just another political leader in Uttar Pradesh."[33]

The people had been demanding greater compensation and some people were not ready to give up their land which was taken away for the Yamuna Expressway and Tappal Township projects of the Jaypee group. Meanwhile, the work of the Expressway was suspended and delayed for a while and a commission was constituted to look after logged complains against the project.[34] The farmers at Jehangarh, an affected village, were very apprehensive. "Yadi je sarkar aai gai, tau kisan phir maaro jayega (if this government comes to power again, then the farmers will be penalized again)", said one farmer. Being a largely Brahmin dominated village; the residents of Jehangarh were mostly BJP voters but decided not to vote for BJP this time. "*Hum vote barbad nai karenge, jo BSP kau harayego tai vote denge, je hatyari sarkar nai aani chahiye phir* (We will not waste our votes, but will vote for whoever can defeat BSP. This murderous government must not come in power again)."[35] The land agitation shaped the attitude of the people that the BSP

[33]Tiwari and Pandey (2013, p. 117).

[34]Ibid.: 118.

[35]Ibid.: 122.

was a murderous government and should not be elected to power again. Therefore they were ready to vote for anyone who was in a position to defeat the BSP.

Meanwhile the entry of Rahul Gandhi in the land agitation and its wide coverage in media politicized the issue. "It started demanding sound bites from the leaders of various political parties. The political parties with eyes firmly fixed on the state assembly election due in May 2012, jumped into politicize this issue. Unable to challenge the BSP government so far on other issues, it had found a subject to embarrass it. The Congress hoped to use it in its struggle to displace the BSP and BJP, aware of the need to reverse the decline of the party to single-digit seats in 2009, was quick to undertake fasts and court arrests."[36] Since the news appeared across the media, politicians across political parties rushed to take advantage.

However, the agitation for greater compensation led to a tussle between the local administration and villagers and claimed the lives of three persons in the village. The sudden eruption of Bhatta Parasul land stir marked a change in the political discourse before election when all political parties extended their support to the people and criticized the state government for being brutal towards the farmers in the state. The issue suddenly attracted the attention of political parties who had a political stake in the state when three people died and around hundred injured in a clash with police on May 7, 2011 in Bhatta Parsual (Greater Noida). The issue was given a fresh leash of life brought to life when Rahul Gandhi visit the Bhatta Parsual village and listened to the people's grievances and promised to take appropriate action over the issue and which was covered by the media. However, the media closely followed Rahul Gandhi and the politics on land agitation and the political consequences thereafter.

Rahul Gandhi got his first political assignment prior to the Uttar Pradesh Assembly election in the form of a land agitation of Bhatta Prasual and since then, he often raised the issue of appeasement of farmers by the BSP government across his political rallies during election campaigns. The issue was discussed in the media as being a national issue from the Parliament to the streets when Rahul Gandhi joined the land stirs in the state as well. However, the issue gradually lost its both media coverage and political interest. Even Rahul Gandhi never visited the village after his first visit. The issue soon lost its political prominance which was found in the early phase of the agitation and at the end of Uttar Pradesh Assembly election campaign the media concluded the issue of *Bhatta Parsual and Tappal incident were lost in election fervour.*[37] The issue lost its relevance and politicians were also not bothered and rather divided the farmers. The issue lost its broader base when once it was considered to be having the potential of becoming a national issue, as the "media played a great role in making this (Bhatta Prasual) issue of national importance."[38]

---

[36]Kumar (2011).

[37]See for details in news under headline 'चुनावी शोर में टप्पल, भट्टा पारसौल कांड गौण' on 25.2.2012 on Rastrya Sahara.

[38]Interview with Akash Singh. He is an IT professional in Delhi based firm and settled in Delhi.

The Bhatta Prasual land stirs generated a fresh wave of agitation against the BSP government. The participation of Rahul Gandhi in the land stirs gave a ray of hope to the villagers who had been fighting against the state government's land acquisition policy. But the passive approaches of politicians against the land agitation of villagers resulted in the fading of the news as the media itself acknowledged."[39]

The people of Bhatta Parsual and Tappal were quite angry with Rahul Gandhi more than any other politicians. This was because the people believed that he was in power and could change their destiny but he never looked at them while they waited for him—evident from the media projection of the issue in the following headline "Bhatta villagers await Rahul's return."[40] Rahul Gandhi for a moment emerged as a crusader for villagers and politicians predicted that he could influence the voting patterns in the region. "There is no doubt that Rahulji restored peace in the two villages. Had he not intervened, both villages would have turned to ashes," said Mukhat Lal Sharma, a Bhatta resident, who was among those injured during the May 7, 2011 clash. "However, after his foot march in July last year (2011), not a single politician or administrative official has come to enquire about the village. The scars of the violence remain in the village. The Congress certainly won our hearts, but it remains to be seen if Rahul will return to campaign", he further added. "While Rahul Gandhi enthused the villagers with his stay and foot march, however, villagers feel that while Rahul is sincere, and his party does not really come up to par. It will make a big difference if Rahul revisits the village", said Lakhim Chand Sharma, another victim of the May 7, 2011 clash.

When Rahul Gandhi visited Bhatta Prasual during the land stirs, he accused the BSP government for acquiring land for the Yamuna Expressway Project without giving proper compensation. The visit of Rahul Gandhi was criticized by the other political parties and they categorically said that the visit had been politically motivated. When Rahul Gandhi sneaked into Bhatta-Parsaul to confront the Mayawati government over land acquisition, it was considered the perfect prelude to the 2012 Uttar Pradesh Assembly elections.[41] In a way, Congress had sounded the poll bugle from the twin Jat majority villages. Before the polls, the BSP had remained unpopular in the two villages, barring the support of Dalit Jatavs constituting about 10% of the electorate. The initial response of the villagers towards the Congress Party seemed tepid and villagers said, for them it was a choice between a rock and a hard place. Even in their unity against the incumbent BSP government, the Jat and other castes could be divided between the Congress and the JD(U) with the latter having fielded farmer leader Manveer Tewatia as its candidate from Jewar. Jewar comprising Dankaur and Jewar blocks, comprised of around

---

[39]Yamuna Expressway is a 6-lane (extendable to 8 lanes), 165 km long, controlled-access expressway, connecting Greater Noida with Agra. It is longest six lane expressway stretch in India. See for details, http://en.wikipedia.org/wiki/Yamuna_Expressway, accessed April 3, 2012.

[40]See details in news under headline 'Bhatta villagers awit Rahul's return', http://timesofindia.indiatimes.com/city/noida/bhatta-villagers-await-rahuls-return/articleshow/11692770.cms, accessed January 31, 2012.

[41]Tiwari and Pandey (2013, pp. 121–122).

2.87 lakh voters. The Congress candidate, Thakur Dhirendra Singh, the man who drove Rahul to Bhatta Parsual on May 11, 2011 and the SP candidate Bijendar Bhati began campaigning in the two villages after the nominations.[42] However, the peoples' anguish against Rahul Gandhi paid the Congress candidate in the elections and the Congress Party lost the constituency to RLD in Khair (SC) where village Tappal is located.[43]

Although, the land acquisition stir enhanced the compensation and resettlement provisions, the mediated politics can be understand in terms of 'before' and 'after' Rahul Gandhi joined the land stir and visited the village. Media played an important role in reshaping the issue. In the '*before*' context, the issue remained as an agitation, lying in isolation. But when media reported Rahul Gandhi's visit to the village it became a highly sensitive and political issue which saw a national debate on issues of land acquisition and development that called for reforms in the land acquisition act.

The issues brought to light by this agitation eventually led to a mandate by the people and media in changing the Land Acquisition Act 1894. The historic Act replaced the 120-year-old legislation of Land Acquisition Act 1894 and was passed by both Houses of Parliament with overwhelming majority during the Monsoon session in 2013. The law replaced the Land Acquisition Act of 1894 by establishing new rules for compensation as well as resettlement and rehabilitation and would be in effect from January 1, 2014.[44] The issue was discussed in both print and visual media in various capacities and interpreted by many as a 'need to change' the old Land Acquisition Act of 1894 as well as empower the farmer.

The new Land Acquisition Act (Rehabilitation and Resettlement), 2013 can be understood as an extension of mediated politics wherein media shapes the issue and plays an instrumental role in replacing the old law with new law. The issues were discussed from all political perspectives before taking a new shape. Thus, the media played an important responsible role for ushering better land acquisition policy in the country. The social space of Bhatta Parsual will always be known for its contribution in the shaping of the land policy of the country. In Hindi media the land stirs of Bhatta Prasual were represented as a forgotten issue and the issue was unable to find takers even during election time while the English media represented the issue in grievances, orders and actions followed by police and administrative provisions. This land agitation of Bhatta Prasual was interpreted in the following manner—involvement of Rahul Gandhi, centre–state relation, insensitive attitude of BSP, land acquisition policy and its reforms.

Land stirs are common but the agitation in the village of Bhatta Prasual (Greater Noida) was something unique because it attracted the attention of the media and

---

[42]See details in news under headline 'Bhatta villagers awit Rahuls return', http://timesofindia.indiatimes.com/city/noida/bhatta-villagers-await-rahuls-return/articleshow/11692770.cms, accessed January 31 2012.

[43]Tiwari and Pandey (2013, p. 122).

[44]http://www.livelaw.in/land-acquisition-act-2013-comes-into-effect-from-today/, accessed January 1 2014.

politicians including Rahul Gandhi and others as well. The issue in fact started a fresh political discourse on centre–state relationships. Land is a state subject and the state government has the authority to make laws regarding the issue. But, in case of discrepancies the Centre's rule will prevail over the state rule under some conditions. The Bhatta Prasual land stirs widened the gap between the centre–state with both governments accusing each other and failing to take the issue seriously. The manner in which incident was reported across the media shows the insensitive approach of BSP government towards poor farmers. However the best part of the land agitation was that it shaped the new Land Acquisition Bill (Resettlement and Rehabilitation), 2013 passed in the Parliament with effect from January 1, 2014.

6.5 Governance and the Politics of NRHM Scam

The mismanagement of National Rural Health Mission[45] (NRHM) was brought to the surface by the media a few months before the scheduled Uttar Pradesh Assembly election and new revelations were made the NRHM scam a popular issue in the media and political parties. The NRHM scam was discussed at length in the media where opposition parties criticized the then Uttar Pradesh government of mismanagement of funds and poor governance. The scam had by now been responsible for the death of three medical officers in the state including Dr. Sachan (Image 6.12).[46] The killing of three medical officers made the issue a very high profile case and making agencies take swift action. However, Anna Hazare's call over zero tolerance on corruption also pressurized the authorities to take action. The Allahabad High Court made an intervention after irregularities were reported and the court asked the concerned agencies to expedite the enquiry process. The Comptroller and Auditor General's (CAG) team looking into the NRHM alleged financial irregularities reportedly accused senior bureaucrats, particularly district magistrates and health officials, for not supporting them in the audit, which was widely publicized in the media especially with Mukesh P Singh, Principal Auditor General (PAG) letter to the Chief Secretary of the state Anoop Mishra in this regard.[47]

[45]The NRHM was centrally sponsored health programme an initiative undertaken by the Government of India to address the health needs of underserved rural areas. The NRHM was launched in April 2005 by Prime Minister Dr. Manmohan Singh to address the health issues in eighteen states including Uttar Pradesh. The sanctity of NRHM was purely holistic but over period the health programme become centre of loot purposefully rather to give health care to the poor rural peoples.

[46]Tiwari and Pandey (2013, p. 153).

[47]See detail under news headlines 'Is Uttar Pradesh serious about NRHM audit?' on 22.10.2011 on TOI, http://timesofindia.indiatimes.com/city/lucknow/Is-UP-serious-about-NRHM-audit/articleshow/10448044.cms, accessed October 22 2011.

Image 6.12 Politics on death. *Source* Aaj Tak, October 13, 2011(Dr. Sachan)

Allegation and counterallegations along political lines were made in the media. The media widely reported the survey conducted by Uttar Pradesh Youth Congress Party that found number of irregularities and mismatched data in the Janani Suraksha Yojana (JSY), a flagship programme under NRHM which was meant to save pregnant women from untimely death. But the scheme had not been able to reach out to people who were really in need, as per a field survey conducted by the Uttar Pradesh Youth Congress. According to the media reports, Ajay Srivastava, state president, Uttar Pradesh Youth Congress stated that, "there has been a lot of fudging in the data". He cited the example of Mira, a native of Shravasti. As per the records, Mira delivered a baby twice in six months. "She confirmed getting the amount for the first delivery but said that she never got pregnant thereafter. Thus, there was no question of applying under JSY," said Srivastava. Citing another case from Gosainganj area of Lucknow, he said that records showed that Anita delivered her baby on June 22, 2011, and Rs 1400 was disbursed to her. But she denied being pregnant at that time.[48] The enquiry progressed and preliminary indications led to the involvement of ministers who was later expelled from the Cabinet and subsequently the Central Bureau of Investigation (CBI) made raids.[49]

The revelation of corruption charges in the NRHM scheme and its discussion in media pushed Mayawati to take action against concerned minister to avoid further criticism. One of the senior BJP leaders Kalraj Mishra alleged the then Chief Minister (CM) of misgovernance and mismanagement of government funds. The media in particular reiterated his statements that while the CM was removing her corrupt ministers, the people of the state would remove her in the state elections.

[48]See detail in news under headline 'NRHM funds embezzled; Congress', on 16.11.2011 on TOI, http://timesofindia.indiatimes.com/city/lucknow/NRHM-fundsembezzledCongress/articleshow/10747950.cms, accessed November 16, 2011.

[49]See detail in news under headline 'भ्रष्टाचारियों की मुखिया हैं मायावती' on 1.1.2012 on Aaj Tak.

NRHM is a centrally sponsored scheme and one of the flagship programmes of UPA I and II, in fact, the NRHM issue shaped the formation of UPA II during 2009 Parliamentary election when both MGNREGA[50] and NRHM ensured the victory of Congress Party. NRHM scam was interpreted as a case of extreme corruption charges against Mayawati, and Babu Singh Kushwaha for swindling of public funds. The issue became very sensitive particularly after Anna Hazare's agitation against corruption and zero tolerance for corruption in the public domain with media being proactive with reporting of cases of corruption and misgovernance. This created a popular sentiment among people who through media expressed their desire for clean and effective governments that would deliver promises made during election campaigns. Government that would be more accountable brings about development effectively. Thus, "the NRHM scam proved to be BSP's waterloo, tainting its image with massive corruption charges, and making it the target of public anger, which finally threw it out of power."[51] The issue played an influential role in shaping the anti-incumbency wave against the BSP in the state when the then CM was forced to sack more than twenty six ministers of her cabinet who were charged with corruption by the Lokayukta.[52]

6.6 Media and the Politics of Anti-incumbency Wave

In politics, anti-incumbency is a political phenomenon that all incumbent governments are faced with. The anti-incumbency is indeed a political stage when political parties in opposition political parties put pressure on the governing political party alleging that the government is a failure and has not responded to the people's grievances. Anti-incumbency is a political fear when the governing political party makes all possible attempts to play down its effects. The media in this case played an important role emphasizing the allegations and counterallegations of the political parties in opposition against the incumbent state government.

[50]The National Rural Employment Guarantee Act 2005 also known as the "Mahatma Gandhi National Rural Employment Guarantee Act", and abbreviated to MGNREGA, is aims to guarantee the right to work and ensure livelihood security in rural areas by providing at least 100 days of guaranteed wage employment in a financial year to every household whose adult members volunteer to do unskilled manual work. For more details see. http://en.wikipedia.org/wiki/National_Rural_Employment_Guarantee_Act_2005 and http://nrega.nic.in/, accessed October 12, 2013.

[51]Tiwari and Pandey (2013, p. 154).

[52]See detail in news under headline 'Mayawati sacks minister number 26' on 19.01.2012 on TOI, http://timesofindia.indiatimes.com/city/lucknow/mayawati-sacks-minister-number-26/articleshow/11547973.cms, accessed January 19, 2012.

Image 6.13 Politics of opposition. *Source* Aaj Tak, February 12, 2012 (Mulayam Singh Yadav)

6.6.1 Mapping the Anti-incumbency Political Discourse Between SP and BSP

The opposition political parties accused the BSP leadership of failing on several fronts. Mulayam Singh Yadav (Image 6.13) openly accused the incumbent state government of corruption and mismanagement of fund. He even added that the administrative machinery too had failed under the incumbent state government's regime.[53]

Shivpal Yadav in an interview with the media hurled vitriolic accusations against the incumbent government and its administration holding it solely responsible for the NRHM scam and demanded a CBI probe.

The media reported some sensitive news as *six died after brain fever in Uttar Pradesh*[54] and discussed the issue as an administrative failure. Medical facilities collapsed as the encephalitis epidemic continued to rage (Image 6.14).

The anti-incumbency wave grew stronger when the people of Hathia village of Azamgarh district suddenly created uproar and said *No bridge then no vote* on river Tamsa which separated Azamgarh town from village Hathia (Image 6.15). Similarly, the people of Besaipur village also said, *No road, no vote* and the people of both the villages threatened to boycott the election process.[55]

Similarly, in the media, the congress was seen as criticizing the incumbent government of misgovernance and mismanagement. For example, in a political rally at Giddor village of Badaun district Digvijay Singh openly accused the

[53]See news in detail under headline 'Maya become richest CM in shortest time: MSY' on 19.1.2012 on TOI, http://timesofindia.indiatimes.com/city/lucknow/maya-became-richest-cm-in-shortest-time-mulayam/articleshow/11547697.cms, accessed January 19, 2011.

[54]See for details in news under headline 'यूपी में दमिागी बुखार से **6** और लोगों की मौत' on 14.10.2011 on Aaj Tak.

[55]See for details in news under headline 'पुल नहीं तो वोट नहीं' on 3.12.2011 on Aaj Tak.

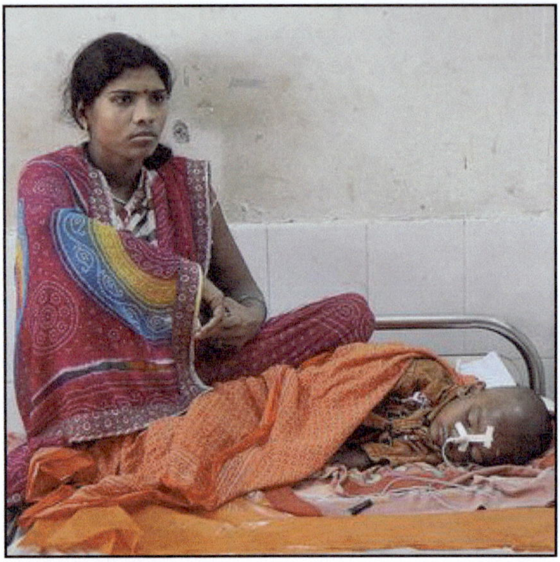

Image 6.14 Politics of cause and concern. *Source* The Hindu, October 29, 2011 (An encephalitis patient at BRD Hospital in Gorakhpur—*Image* Subir Roy)

Image 6.15 Politics of agitation. *Source* Aaj Tak, December 3, 2011 (peoples of Hathia village)

administration violating people's right and questioned the law and order situation in the state. He also questioned the BSP being the champion of the underprivileged and marginalised when it was trying to lure other castes "Hathi Nahi Ganesh Hai, Brahma Vishnu Aur Mahesh Hai" for the purpose of vote bank politics. Similarly, Sachin Pilot in a rally at Agra claimed that the *Congress will form the government in Uttar Pradesh*[56] and the Congress government would ensure all round development. He emphasized on the expansion of broadband technology in the state and further added that during BSP's rule the poverty rate increased. Akhilesh Yadav in an election rally at Singraully as covered in media urged voters to critically analyse the achievements of the BSP government and added that in the past four-and-a-half years, the state was pushed back economically, accusing the government failing to serve the interests of poor and farmers.

6.6.2 Resisting the Anti-incumbency Politics—Representations of BSP in Media

BSP in their counterallegations in media asked the Congress Mayawati criticized the Congress not to overdramatize. Rather Congress should focus on the Congress Party ruled states instead of laughing at Uttar Pradesh. She denied the arguments made by Congress regarding backwardness, poverty, exploitation of farmers and advised the Congress to check the prices of vegetables and other commodities and to not undermine the people of Uttar Pradesh. She completely rejected the arguments made by opposition parties that she had given undue advantages to other upper castes for keeping intact other caste, particularly upper caste based lobby within BSP. She, however, responded by saying that the BSP believes in "Sarvajan Sukhay, Sarvajan Hitay" (Happiness to Everyone and Favour to Everyone) and that they (other castes and upper castes) too believed in the same ideology. Therefore, they are with BSP. In media, Satish Mishra, a popular and powerful upper caste leader of BSP, praised Mayawati as the *deity of law and justice*[57] and appealed to upper caste voters not to fall in the trap of opposition parties and shouted like *Brahmins will blow the conch and the elephant will march on*. The media widely reported that apart from Brahmin voters, BSP was trying to woo Muslims and other upper caste voters to ensure the BSP winning in the forthcoming elections. In fact the media widely reported that the BSP to ensure victory in 2012 responded to add any allegations made and even sacked 26 ministers in 6 months after sensing the anti-corruption mood among the voters in the state. Mayawati was reported in the media of having conducted a secret survey of BSP's sitting MLAs and all the relevant information was collected from available sources including police and intelligence reports, and the party denied giving party tickets to more than hundred sitting MLAs in the election. The action taken by the party was considered as part of corrective measures while BSP leaders found "two benefits in party's move: one, they have something in hand to fight anti-incumbency and counter their rivals on

[56]See for details in news under headlines 'यूपी में बनेगी कांग्रेस की सरकार' on 17.12.2011 on Aaj Tak.

[57]See for details in news under headline 'ऊंची जाति के गरीबों को मिले आरक्षण: मायावती' on 13.11.2011 on Aaj Tak.

the issue of corruption; two, by listening to the opinion of BSP coordinators, the party had given a message to the cadre that party organization is more important. However, BSP leaderships also argue that anti-incumbency is not only against the government but also against the sitting MLA's hence new candidates being fielded in place of those denied tickets will have a better chance to win."[58]

The regime of the BSP was portrayed in the media in negative manner where people showed their anguish for lack of infrastructure (road, bridge, etc.) that also reflected bad governance.

6.7 Conclusion

During the Uttar Pradesh Assembly elections, there were a number of development issues raised and discussed but only a few of them were acknowledged and reported in the media. The media primarily reported Centre–State relationship, flagship programme, bad governance and caste politics. The media shaped the development discourse in a way to rationalize the importance of "heartland" that is Uttar Pradesh in the India politics and that development was no longer a rational issue but equally important local one. The national parties like Congress Party and BJP seemed in media to address the issue very seriously while regional political parties (SP and BSP) were not serious and instead resorted to making allegations. The Congress Party alleged that the grants sent by the Central Government over time were either unutilized or mismanaged. Congress and BJP leaderships targeted the incumbent government in media over issues of development and governance. Thus, making development and governance as key issues that were no longer limited to national debate and also at the regional level. It also highlighted a shift from narrow caste and religion-based politics to more progressive and secular development and governance-based politics.

The next chapter is conclusion wherein research questions and objectives were specifically addressed on the basis of preceding chapters. These chapters are supposed to address the specific study on mediascapes and the state that how duo works together or separately built an image(s) either for political or institutional practices. This study considered Uttar Pradesh Assembly election-2012, as case study to understand the construction of an image(s) that how such image(s) become contentious over time.

---

[58]See for details under news headline 'No tolerance or pure politics?' on 8.1.2012 on TOI, http://timesofindia.indiatimes.com/home/stoi/deep-focus/-No-tolerance-or-pure-politics/articleshow/11408241.cms?, accessed January 8, 2012.

References

Kumar A (2011) The battle for land: unaddressed issues. EPW 46(25) (June 18, commentary page)
Kumar S et al (2012) Sixteen assembly elections in Uttar Pradesh. EPW 47(14):80–86 (April 7, 2012)
Narayan B (2012) Uttar Pradesh's renouncer king. EPW 47(6):31–32 (February 11, Mumbai)
Tiwari M, Pandey R (2013) Battleground UP: politics in the land of Ram. Tranquebar Press, Chennai

Chapter 7
Post Script

Abstract The study enlighten on mediascapes and the state with respect UP Assembly election-2012 with the help of eight issues. Both mediascapes and the state mapped UP as "Heartland" of Indian politics. The political discourse was mapped both negative politics (caste and communal) and positive politics (development and backwardness) as well as to address the dichotomy of centre versus the periphery or national versus regional or development versus backwardness.

Keywords Public · Private · Centre–state · Social space

Media has emerged as a powerful institution over the years and has captured peoples' attention in various ways. In the current scenario, people have become dependent on media to access and absorb all kinds of information, including spatial information. Apart from the information relayed by media, political news has become one of the main constituents of media. During the UP Assembly elections in 2012, the media played an instrumental role and shaped and reshaped the political discourse wherein national political views were given an edge over regional political views.

While the media represented the political views of respective political leaderships in the form of news, actual textual interpretation was further shaped by visual media into pictorial images and those images manifested into symbols to represent the space as well. During the UP Assembly elections, many images were produced and reproduced in which many of them were portrayed across media in various capacities. Media used these images and represented the state (UP) during the elections in multiple ways.

This study has concluded the relationship between media and the state with reference to UP Assembly election in 2012 and discussed under four research areas where this study made stride.

S. Moinuddin, *Mediascape and The State*, Springer Geography,
DOI 10.1007/978-3-319-51932-6_7

7.1 The Nature of Indian Media

Media developed in order to provide communication facilities effectively and efficiently and to become an intervening junction between the "public" and "private" spaces. In India, urban population primarily views and consumes privately owned media, while Doordarshan, the state-owned television channel, remains popular primarily in the rural areas because of the social, cultural and economic gaps that restrict people's viewing choices. However, print media is completely run by private players with government subsidy. In terms of circulation, Hindi and vernacular newspapers have outnumbered the English newspapers, wherein the latter is restricted in its circulation predominantly in the urban areas. Though it appears that there is battle for news between TV and print, it is incorrect. TV and print media together account for a large chunk of the money that the business of news generates but the internet and mobile phones are fast emerging as serious contenders, at least in terms of volume.

The Indian news market provides equal opportunities to both vernacular and English language media to grow horizontally and vertically. Amongst the top ten daily newspapers in the country, five are in Hindi, four in vernacular languages and one in English. The Indian press has been categorized into two sections—English and the vernacular press. The former is seen as being "national," while the latter as being the 'regional' press. Overall, the English press has a higher status than the vernacular press and similarly, the national press enjoys a higher status than the regional press.[1] However, the circulation crossed more than 320 million while in terms of registered newspapers, Uttar Pradesh and Delhi emerged as having the largest circulation in India. The radio service covers almost 99.16% of the population while it covered around 92% of the total area of the country so far. The expansion of television service in both rural and urban spheres is gradually increasing but in the rural areas, the rate of growth is relatively higher than that in the urban spaces in India. The rate of growth of social media is exemplary and the major use of social media comprises the usage of email and social networking. Young men and college-going students are the main users of social media in the country. Around 35% of active users of social media are found in top eight metro cities, while 11% are found in non-metro cities in India. Activities such as email, social networking and chatting are indulged in the most through mobile phone Internet facilities.

However, market competition and the effects of globalization forced newspaper organizations to set up local editions to tap the local market as well. Almost all newspapers in the country more or less started their local editions and expanded the newspaper business in the country. Meanwhile, the estimated television viewership in India has been projected as 200 million by the end of 2017. The preference of cable, satellite and DTH connections over signal receivers also increased manifolds in both rural and urban India. The digitisation phase is expected to be over in the

[1]Stahlberg (2013), p. 13.

next 2 years throughout the country. DD is a national channel and is assumed to reflect the country's position on various issues, while private channels promote their commercial interests through the nature of news that is covered and represented in terms of news content. The rationality and neutrality of both state-owned and privately owned TV channels has been questioned regarding the coverage, representation of news and its content. Based on Bennett (2007), the following trends have been observed in Indian media as well:

1. "Information trends 1; Less Public Service and Alternative Programming"
2. "Information Trend 2: Internal Corporate News Censorship and Self-Advertising as News"
3. "Information Trend 3: More infotainment and Soft News"
4. "Information Trend 4: Generic News"
5. "Information Trend 5: Branding the News".

7.2 How Media Constructs UP as the "Heartland" of Indian Politics and Polity

UP can be seen as being the heartland of Indian politics in multiple ways and has been through five such distinct, though overlapping, constructions through modern history—UP as a "colonial heartland," as a "nationalist heartland," as a "Hindu heartland," as a "Muslim heartland" and finally as a "postcolonial heartland."[2] Media often discusses the political strength of UP in terms of centre–state power correlation. The first Assembly elections were held in 1952 in the state when the Congress Party emerged victorious and Govind Ballabh Pant assumed the charge of Chief Minister of the state in independent India. Subsequently, in the Central Government, the Congress Party under the leadership of Jawaharlal Nehru assumed the office as the first Prime Minister of independent India. The Congress Party was in power till April 3, 1967 at both the Centre and the state when Charan Singh became the first non-Congress Chief Minister of the state. His rule in the state was considered an exception as the Congress Party was not in power in the state but was in power in the Central government. After the period of Emergency, when elections were held, the Congress Party lost the elections and remained out of power for nearly three years. Subsequently, in that period, Ram Naresh Yadav and Banarsi Das of Janata Party ruled in the state, while in the Centre, Morarji Desai and Charan Singh of the Janata Party were in power.

During the 1990s, the emergence of regional political parties was witnessed and coalition government was formed when Mulayam Singh Yadav assumed the office in the state and at the Centre, the National Front came to power under the leadership of V.P. Singh with the support of BJP. But the government fell in both places—in the state and at the Centre. Thereafter the United Front, NDA, UPA I & II were

[2]Kudisya (2006).

formed over time under different coalition governments, and was backed by the ruling political party in UP as well.

Media emphasizes and covers news related to particular constituencies that emerge as power centres in Indian politics, as they are mostly associated with prospective Prime Ministerial candidates. UP has produced 10 Prime Ministers out of 15 so far and comprises 80 Lok Sabha seats, which is 15% of the total Lok Sabha seats. As a result, places like Amethi, Rae Bareilly, Mainpuri, Rampur, and Varanasi have become the key constituencies in Lok Sabha elections and reiterate the predominance of Uttar Pradesh in National politics. It also demonstrates the way in which media, through its wide coverage, has transformed these regional constituencies into major constituencies that have come to symbolize particular parties and their prospective Prime Ministerial candidates.

During the 1980s and 1990s, the state witnessed the emergence of Dalit politics under the leadership of Kanshi Ram and Mayawati, both of whom shaped the Dalit movement in the state. The face of Dalit politics was quite radical during the 1980s and 1990s, as it openly addressed the oppression of the lower and backward caste people. However, during the year 2007, Dalit politics in the state witnessed a shift from Bahujan Samaj to 'Sarva Jan Hitaye'—an inclusion of all castes, particularly those who have been marginalized in their empowerment and development. It is important to note that though Dalit politics was initiated in UP, it became a major national issue that addressed the empowerment and development of marginalized communities, specifically those based on caste.

Media shapes political discourses and the coverage of Babri Masjid–Ram Mandir (BM–RM) issue in UP, followed by the subsequent Ram Janambhumi Movement (RJM) across the nation brought back the issue of secularism and secular politics as a major political concern at the national level. This, once again, reiterates the position of UP as a heartland and the way in which it predominates national politics in terms of the manner in which regional issues assume national importance. Both the Dalit and Babri Masjid–Ram Mandir issues have become national issues over time while they were earlier restricted to the state and the role played by the media in doing so is crucial.

7.3 How Political News Is Shaping and Reshaping the Social Space of Uttar Pradesh

The caste politics witnessed during the UP Assembly elections in 2012 is illuminating as almost all political parties had a political blue print to lure upper caste voters. BSP experimented with the caste configuration in the form of social engineering in 2007, which was quite successful, when they brought Upper caste and Dalits people on the same political platform under its leadership. SP on the other hand followed the Muslim-Yadav (MY) political formation. Meanwhile, both national parties—BJP and Congress—also reformulated their political plan and lured the forward class and marginal people within their political formation.

However, during elections the relationship between caste, religion and politics was evident and almost all political parties had pushed such policies to lure people along the lines of caste and religion. Cutting across political parties and their ideological positions, the current UP Assembly elections in 2012 witnessed a merging of multiple caste groups under a broader social structure that became symbolic of the politics of inclusion. For example, Mayawati included the poorer upper castes and discussed the need for reservation for them—Deprived forward caste people must get reservation.[3] She further described the Brahmins as being the "exploited lot."[4] The caste politics in UP has undergone a shift with political parties luring upper caste and the role of upper caste was evident during the 2007 UP Assembly elections.

The minority discourse, primarily the issue of reservation for minorities, was reshaped by the media to indicate only Muslims though it includes other communities as well. This mediated discourse was shaped and interpreted as Muslim reservation, whereas Muslims are only one of the many communities that constitute minorities. Both print and visual media have portrayed minorities as Muslims, irrespective of the news content. Except *NDTV 24*7* and *Rashtrya Sahara*, all media houses diluted the issue. Around 73% news content in *Aaj Tak* and around 83% news content in *Zee News* on reservation for minorities were interpreted as an issue regarding reservation for Muslims instead of reservation for minorities. Amongst the English news channels, *CNN-IBN* discussed the issue as that of Muslims. Interestingly, *DD*, the state-sponsored news channel, followed the issue and interpreted the minorities as being Muslims. Amongst the newspapers, *The Times of India* and *The Hindu* interpreted the issue in a similar vein. Thus, media reshaped the discourse and constructed the issue as a Muslim issue and popularized the notion of minorities being synonymous with Muslims. Of course, the issue was discussed in the light of vote bank politics during the election period when Congress Party brought in the provision of 4.5% reservation for minorities under OBCs quota.

During the UP assembly elections, the media discussed the issues that shaped and reshaped the image politics around UP. In doing so, media discussed four broad issues and portrayed the state government of UP in a negative manner and brought to the forefront notions of corruption, agitation, frustration and the deteriorating law and order conditions in the state. The following issues were brought up—NRHM scam; Bhatta Prasual land stirs; Anti-incumbency wave and Rahul Gandhi: A tireless campaigner.

Thus the image of the BSP government that was created by the media was that it failed to deliver the expectations of people. The issue was largely seen as a case of corruption and caste politics with the expulsion of a particular leader from the party and his later inclusion in the BJP. Media represented the incident as the leaders

---

[3]See in details in news under headline 'Reservation to poor forward caste people' on November 13, 2011 on Zee news.

[4]See in detail in news under the headline 'Mayawati describes Brahmins as "exploited lot" on November 14, 2011 on The Hindu, [http://www.thehindu.com/news/national/mayawati-describes-brahmins-as-exploited-lot/article2624905.ece, accessed November 11, 2011].

being unable to run the government, being corrupt and all this ultimately leading to a lack of development in the State.

Another incident, the Bhatta Prasual land stir issue was represented in the media in four ways: the involvement of Rahul Gandhi; centre–state relation; insensitive attitude of BSP against poor people and apathy against land policy. Land stirs are common but the agitation in the village of Bhatta Prasual (Greater Noida) had just taken place before the Assembly elections, and therefore it attracted the attention of media and politicians including Rahul Gandhi and others. The villagers claimed that the authority forcibly took their agricultural land without giving compensation and putting in place a proper rehabilitation policy. The people of Bhatta Prasual opposed the authority from grabbing the land forcibly and the local administration called the police to restore the situation which resulted in the death of three people. The issue surfaced in the media when Rahul Gandhi visited the village to console the people and criticized the BSP state government. The issue escalated the conflict between the Centre and the state, whereas land is a state matter and the state government has rights to make rules regarding it. Therefore, BSP registered their dissent against Congress Party lead UPA II government. Mayawati counter attacked Rahul Gandhi and UPA II and other major political parties for not preparing a pro-farmer land policy. The Bhatta Prasual land stir shaped the new land policy of the country and replaced the historic 120-year-old legislation of the Land Acquisition Act, 1894 with a new land policy when it was passed by both the Houses of Parliament with an overwhelming majority during the Monsoon session in the year 2013 and the law became effective from January 1, 2014.

Media discussed the role of Rahul Gandhi during UP election with specific objectives and the whole issue was discussed in three ways—Rahul Gandhi as a leader of the Congress Party and its prospective Prime Ministerial Candidate; UP Assembly elections as a prelude to the General elections in 2014 and Developmental politics. The UP Assembly elections 2012 were of course a prelude for Rahul Gandhi as the Congress party's Prime Ministerial candidate as he was portrayed as taking command of UP Assembly elections and addressed 211 rallies in a span of forty-two days, covering 375 out of 403 assembly constituencies since his first election rally from Phulpur in Allahabad on 14 November 2011 to 29 February 2012 in Bareilly. He conducted 18 road shows in cities like Lucknow, Kanpur and Bareilly amongst others.

The UP Assembly elections 2012 were discussed in the media with an eye on the Central government and the future of UPA II government and possible coalitions emerging that would form the government in the Centre in 2014. In fact, all these discussions made UP elections a decisive political election, possibly predicting the future government that may come to power at the Centre.

In his election rallies, Rahul Gandhi addressed issues of backwardness and the need for the implementation of a systematic growth plan in a focused manner and came up with a road map for UP's development, the "Vision Plan 2020." At Barabanki, Rahul Gandhi claimed that UP will become a developed state in 5 years and UP can become like Maharashtra and Delhi, which indicated that the only agenda for the Congress Party was development.

The anti-incumbency issue was discussed in the media in many ways including the issue of governance and development. Everyday media came up with news that showed the ways in which the State government had failed in delivering good governance and the development of UP. Media coverage included reportage and news on the death of people due to brain fever in UP and people demanding development—"no bridge no vote" and No road no vote. The media in this particular election extensively discussed governance and development as the major concerns of almost all political parties and perhaps for the first time shifted focus away from caste and communal politics.

7.4 How Political News Constructs the Image of Uttar Pradesh

Media constructed UP as the heartland of Indian politics and polity and made the UP elections in 2012 a curtain raiser for the Parliamentary elections in 2014. Media discussed all potential issues during the election and gave UP political importance not seen in the case of other states. The image of UP as the single largest state and its influential political stature was challenged by Mayawati when she proposed its division to make four new states, namely Bundelkhand, Paschim Pradesh, Awadh Pradesh and Purvanchal. The proposal was not successful and did not gain any additional coverage in media as well.

Earlier media created an image of UP as a state ridden with caste and communal politics, where vote bank politics and social engineering predominate but interestingly, in this election governance and development became the major concerns as portrayed by the media.

References

Bennett WL (2007) News: the politics of illusion. Pearson and Longman, New York

Kudaisya G (2006) Region, nation, "heartland": Uttar Pradesh in India's body politics. Sage Publications, New Delhi

Stahlberg P (2013) Writing society through media: ethnography of a Hindi daily. Rawat Publications, Jaipur

Appendix A
Uttar Pradesh Assembly Election 2012 Issues

A.1 UP: The Nerve Centre of Politics

The politically crucial state of Uttar Pradesh goes to poll at a time when Mayawati regime is facing a barrage of corruption charges and an anti-incumbency wave, which seems to be getting stronger with every passing day.

A.2 The Anti-incumbency Wave Against BSP

BSP won 206 seats in 2007 and seized power by dethroning Samajwadi Party (SP) government led by Mulayam Singh Yadav on its own. At that time, people were fed up with the deteriorating law and order situation and criminalization of politics in the state.

S. Moinuddin, *Mediascape and The State*, Springer Geography,
DOI 10.1007/978-3-319-51932-6

Voters chose Mayawati as the best possible alternative; breaking barriers of caste and community. However, this time around, things are pretty different. Mayawati may not be the people's favourite anymore. Several ministers are facing inquiries by the Lokayukta making the government vulnerable to opposition attacks. Mayawati has denied tickets to around a hundred sitting MLAs in a desperate attempt to fight off anti-incumbency and to improve her party's image. The BSP is trying to make the best of a bad situation. Party leaders have been challenging rivals by putting forward the 'action' taken by Chief Minister Mayawati against her ministers. BSP leaders feel that this would help Mayawati in fighting anti-incumbency and countering rivals on the issue of corruption; and also help strengthen the party's cadre base by sending a message that their voices are being heard. Moreover, as many of the sacked ministers and MLAs have joined other parties, it has given BSP ammunition to counterattack the naysayers. BSP leaders are of opinion that anti-incumbency is not only against the government but also against the sitting MLAs, and hence new candidates fielded in place of those denied ticket will have a better chance to win. However, anti-incumbency is not the only factor which worries Mayawati. The BSP does have a loyal voter base in Dalits but what had tilted the scales in her favour, last time around, was the support of upper castes, especially the Brahmins. This time around, a resurgent BJP is reclaiming the upper caste voters, thereby making the going tough for the Elephant. As plan B, Mayawati has fielded more Muslims than any other party. But it would a tougher nut to crack given the dedicated wooing of the community by the Samajwadi Party and the Congress.

However, Mayawati does have an ace up her sleeve in form of her plan to divide UP into four smaller states. Mayawati, clearly, faces the challenge of scoring a double century if she wants to retain the reins of power. But whether her Dalit icon status and social engineering formula will be enough to sail through is the big question.

A.3 Rahul Gandhi—A Tireless Campaigner

With Rahul Gandhi taking the centre stage in Congress' battle to reinvigorate itself in Uttar Pradesh, there is optimism that party's 22-year-long political wilderness will finally end. Backed by Congress' "Utho, Jago, Badlo" slogan, Rahul seems to have struck an emotional chord with all sections of the society, especially the youth, who look up to him as a leader who will bring a paradigm shift in governance. To ensure a victory for the Congress, Rahul had been crisscrossing the length and breadth of the state for some time now. Rahul started his election campaign by taking up the emotive issue of farmer's agitation in Bhatta Parsaul in Gautam Buddha Nagar which had witnessed violent clashes between police and farmers. Apart from taking up the "fight for famer rights," the Gandhi scion attacked successive non-Congress state governments for rampant corruption and poor law and order situation. However, the Congress is not just banking upon the personal charisma of Rahul Gandhi and has sewed together a smart electoral alliance with Rashtriya Lok Dal (RLD) of Ajit Singh, which can tilt the scales in its favour in western UP. The grand old party would also expect to reap the political dividend of its decision to carve out a sub-quota for Muslims within the 27% reservation for OBCs, a move credited to Rahul. Moreover, Congress also appears open to a post-poll alliance with the Samajwadi Party. Clearly, the arch light is on the Yuvraj this time around. He had proved his mettle in 2009 Lok Sabha polls when Congress managed to bag 22 seats in UP and its vote share rose to 18.3% from 9%. Political observers say that even if the Congress manages to touch the 100-seat mark, it would be a huge achievement for the party and, in particular, for Rahul Gandhi. Rahul has to win; more importantly 'Rahul factor' needs redemption after Bihar polls.

A.4 NRHM Scam and Kushwaha Fiasco

The Mayawati's government was further pushed to wall when state's former family welfare minister Babu Singh Kushwaha, health minister Anant Kumar Mishra and several others were accused of committing financial irregularities in connection with the centrally sponsored National Rural Health Mission. Kushwaha's name also figured in the alleged murder of three government doctors, which led to a huge uproar as the killings were later found to be linked with the Rs 10,000 crore NRHM scam in the state's Family Welfare and Health Department. Under mysterious circumstance, chief medical officer, Vinod Kumar Arya was killed in a broad daylight on October 2010. This was followed by the killing of CMO Dr. BP Singh outside his home on April 2. The police under pressure from higher ups booked then Deputy CMO YS Sachan for getting the two doctors killed and sent him to jail. Then on June 22 last year, Dr. Sachan was also found dead inside the jail premises. Initially it was dismissed as a suicide by the police but later it was proved that he was killed.

Under pressure, Mayawati ordered a CBI into the matter. She also moved ahead and sacked Kushwaha and Mishra and later expelled them from the party. Kushwaha, a two-time MLC, again courted controversy when it was reported that he was making inroads into Congress. However, his efforts were scuttled by Rahul Gandhi. In an interesting turn of events, the BJP, eyeing the OBC vote bank, later inducted him into the party and decided to field him as its candidate in the coming Assembly polls. After joining BJP, Kushwaha created a stir by claiming threat to his life from BSP colleague Naseemuddin Siddiqui, UP Cabinet Secretary and Principal Secretary (Home). BSP, however, hit back and accused him of indulging in anti-party activities. The induction of corruption-accused Kushwaha caused huge embarrassment to the BJP top brass as the move was opposed by its prominent leaders. Senior leaders LK Advani, Sushma Swaraj and Murli Manohar Joshi expressed their displeasure against Kushwaha's induction and said that the decision had "dented" the anti-corruption agenda of the saffron party. With dissent growing within its ranks, BJP later put Kushwaha's membership on hold and announced that it would neither field him as a candidate nor make him one of its star campaigners in the coming Assembly elections.

A.5 Land Acquisition Stir and Bhatta Parsaul Agitation

Land acquisition has emerged as one of the most important poll issues in Uttar Pradesh. The fire started from Bhatta Parsaul village near Delhi, where the police opened fire of farmers protesting against forceful acquisition of their land. The flames soon spread to other parts of the state much to the chagrin of the state government led by Mayawati. The whole issue acquired political credence after Congress Rahul Gandhi took up the issue as part of his "Mission UP." The state government may have tried to extinguish the fire but considering the continuing problems being faced in acquiring land for development projects, especially after courts took the hardline on the issue, Mayawati may have a lot of explanation to do.

A.6 Division of Uttar Pradesh

Of all her achievements, Mayawati would, expectedly, tried to play the "plan" to split Uttar Pradesh into four parts—Purvanchal, Paschim Pradesh, Avadh Pradesh and Bundelkhand—the most. There may be credible reasons to divide Uttar Pradesh considering the apparent difficulty in governing a state as big as Uttar Pradesh—if Uttar Pradesh would have been an independent county it would be world's fifth largest in terms of population—but Mayawati also knows the appeal of the issue in the state where good governance has always been a utopian dream. The "divide Uttar Pradesh plan" is Mayawati's trump card and she knows it. But more importantly, she gains strength from the fact that opposition is finding it difficult to counter her on the issue.

The Dalit ki beti has cleverly put the ball in the Centre's court hoping that the promise of a "new promised land" would help her tide over the allegations of corruption and nonperformance being levelled against her by the opposition.

A.7 Reservation for Minorities

On December 22, 2011, the Union Cabinet approved reservation for minorities under the Other Backward Castes category and decided to give 4.5% sub-quota to minorities to be carved out from the existing quota of 27% for the OBC. The resurgent Congress under Rahul Gandhi expects to make the most gain from the policy shift in favour of minorities, especially Muslims, who play a decisive role in as many as two dozen constituencies in the state. Predictably, the Opposition especially the BJP denounced move as being opportunistic, saying that it was timed with an eye on the Assembly elections. The Samajwadi Party supports the quota but is wondering how to prevent the Congress from making gains from it. Although, the Election Commission has put the minority quota on hold but it continues to be a rallying point in the election.

A.8 Statues, Memorials, Parks and Elephants

Her detractors call it wasteful expenditure but for Mayawati the gigantic monuments and memorials built are symbols of "Dalit pride." With rows of sandstone elephants and statues of Dalit icons including her own, Mayawati surely knows the emotive value of the endeavour.

Although the Election Commission has covered up her parks and statues and she may have been constantly under attack by the opposition and media, there has been no election in Uttar Pradesh, in recent times, without the commanding presence of the elephants and that stands true for this time as well. What can be called a miracle, Uttar Pradesh joined the club of 7% growth rate. No doubt this is an issue which Mayawati will surely raise in the state elections. This is even more pertinent considering that the state has long been regarded as feudal and caste-driven where fast growth was not a possibility. Notwithstanding what her critics might say, the symbols of so-called Dalit pride might just work in the favour of the BSP supremo. That section of the population who may have felt marginalized by the ruling strata of the society may take this to be a step towards their empowerment.

Note The news prepared by Ritesh K Srivastava/Manisha Singh (Zee News)
Source http://zeenews.india.com/state-elections-2012/up/issues.html/7

Appendix B

Response no…………..   Date & Time…………

Focus Group Discussion

Media and the State: A Geographical Interpretation of Image Politics in U

List of Questions

Youth ☐ Gender☐ Political Activists☐ Professional☐ Religion ☐

1. What do you think is the role of media during elections?

   Yes………..   No………..

2. These issues are as……what are your opinions about this and their relevance in the upcoming elections.
   - UP: The Nerve of Politics
   - Division of UP
   - Statues of Mayawati, Elephant, Parks and Memorial
   - The Anti-incumbency wave against Mayawati
   - Bhatta-Prasual Land Stir
   - NRHM Scam or Kushwaha Fiasco
   - Rahul Gandhi: A tireless campaigner
   - Reservation for Minorities
3. How do you see//understand the role of UP in national politics in India and how media is shaping the overall image of UP through its content.
4. How do you evaluate politics around the proposed division of UP and how media shaping the image politics around this issue.
5. How do you understand the politics around the statues of Mayawati, Elephant and Parks and how do you evaluate the role of media regarding this issue.
6. Do you see any role of media in creating the anti-incumbency wave against Mayawati during UP election?

S. Moinuddin, *Mediascape and The State*, Springer Geography,
DOI 10.1007/978-3-319-51932-6

7. How do you assess the role of media in popularizing the Bhatta Prasual land stirs?
8. How do you assess the role of media in brining the political discussion on NRHM to the forefront during this election?
9. How do you visualize the role of media in order to shape the politics around Rahul Gandhi during the election campaign?
10. How do you assess the role of media to shape the politics around the issue of reservation for minorities during election?
11. Do you feel that media have power to shape the politics of space (locality/region/nation)?
12. Do you feel that news is constructing and giving a new identity to space? To some extent in everyday life is.
13. Do you feel that media emerges as an influential agent which is affecting our day-to-day course of action particularly our politics and its nature?
14. Do you feel that media and political parties are dependent on each other?
15. What do you feel between media and politics (politicians) who is more dependent on whom and why?

Appendix C
General Election Results of 1999, 2004 and 2009

S. No.	Political Party	1999	2004	2009	2014
		Number of seat won (% of voting share)	Number of seat won (% of voting share)	Number of seat won (% of voting share)	Number of seat won (% of voting share)
1.	BJP	29 (27.64)	10 (22.17)	10 (17.50)	71 (42)
2.	BSP	14 (22.08)	19 (24.67)	20 (27.42)	0
3.	SP	12 (24.0)	35 (26.64)	23 (23.26)	5
4.	INC	10 (14.72)	9 (12.04)	21(18.25)	2
5.	Akhil Bhartiya Lok Tantrik Congress	2 (1.51)	N.A	N.A	
6.	Rastrya Lok Dal	2 (2.49)	N.A	5	
7.	Independent	1 (3.62)	N.A	1	
8.	Samajwadi Janta Party (Rastrya)	1(0.46)	N.A	N.A	
9.	Others (including all)	N.A	7 (14.38)	N.A	2 (Apna Dal)

Source http://en.wikipedia.org/wiki/Indian_general_election_1999,2004,2009

S. Moinuddin, *Mediascape and The State*, Springer Geography,
DOI 10.1007/978-3-319-51932-6

Appendix D
Lists of Chief Minister of Uttar Pradesh

S. No.	Name constituency	Term of office		Party	Assembly election
1.	Govind Ballabh Pant Bareilly Municipality	January 26, 1950	May 20, 1952	Indian National Congress	Not yet created
		May 20, 1952	December 27, 1954		First Assembly (1952–57) (1951 election)
2.	Sampurnanand Varanasi City South	December 28, 1954	April 9, 1957		
		April 10, 1957	December 6, 1960		Second Assembly (1957–62) (1957 election)
3.	Chandra Bhanu Gupta Ranikhet South	December 7, 1960	March 14, 1962		
		March 14, 1962	October 1, 1963		Third Assembly (1962–67) (1962 election)
4.	Sucheta Kriplani Mendhwal	October 2, 1963	March 13, 1967		
(3)	Chandra Bhanu Gupta Ranikhet	March 14, 1967	April 2, 1967		Fourth Assembly (1967–68) (1967 election)
5.	Charan Singh	April 3, 1967	February 25, 1968	Bharatiya Kranti Dal	
–	Vacant (President's rule)	February 25, 1968	February 26, 1969	N/A	*Dissolved*

(continued)

S. Moinuddin, *Mediascape and The State*, Springer Geography,
DOI 10.1007/978-3-319-51932-6

(continued)

S. No.	Name constituency	Term of office		Party	Assembly election
(3)	Chandra Bhanu Gupta Ranikhet	February 26, 1969	February 17, 1970	Indian National Congress	Fifth Assembly (1969–74) (1969 election)
(5)	Charan Singh Chhaprauli	February 18, 1970	October 1, 1970	Bharatiya Kranti Dal	
–	Vacant (President's rule)	October 1, 1970	October 18, 1970	N/A	
6.	Tribhuvana Narayana Singh	October 18, 1970	April 3, 1971	Indian National Congress	
7.	Kamlapati Tripathi Chandauli	April 4, 1971	June 12, 1973		
8.	Vacant (President's rule) Hemwati Nandan Bahuguna Bara	June 12, 1973 November 8, 1973	November 8, 1973 March 4, 1974	N/A Indian National Congress	
		March 5, 1974	November 29, 1975		Sixth Assembly (1974–77) (1974 election)
–	Vacant (President's rule)	November 30, 1975	January 21, 1976	N/A	
9.	N.D. Tiwari Kashipur	January 21, 1976	April 30, 1977	Indian National Congress	
–	Vacant (President's rule)	April 30, 1977	June 23, 1977	N/A	*Dissolved*
10.	Ram Naresh Yadav	June 23, 1977	February 27, 1979	Janata Party	Seventh Assembly (1977–80) (1977 election)
11.	Banarsi Das Hapur	February 28, 1979	February 17, 1980		
–	Vacant (President's rule)	February 17, 1980	June 9, 1980	N/A	*Dissolved*

(continued)

(continued)

S. No.	Name constituency	Term of office		Party	Assembly election
12.	V.P. Singh Tindwar	June 9, 1980	July 18, 1982	Indian National Congress	Eighth Assembly (1980–85) (1980 election)
13.	Sripati Mishra	July 19, 1982	August 2, 1984		
(9)	N.D. Tiwari Kashipur	August 3, 1984	March 10, 1985		
		March 11, 1985	September 24, 1985		Ninth Assembly (1985–89) (1985 election)
14.	Vir Bahadur Singh Paniara	September 24, 1985	June 24, 1988		
(9)	N.D. Tiwari	June 25, 1988	December 5, 1989		
15.	Mulayam Singh Yadav Jaswantnagar	December 5, 1989	June 24, 1991	Janata Dal	Tenth Assembly (1989–91) (1989 election)
16.	Kalyan Singh Atrauli	June 24, 1991	December 6, 1992	Bharatiya Janata Party	Eleventh Assembly (1991–92) (1991 election)
–	Vacant (President's rule)	December 6, 1992	December 4, 1993	N/A	*Dissolved*
(15)	Mulayam Singh Yadav Jaswantnagar	December 4, 1993	June 3, 1995	Samajwadi Party	Twelfth Assembly (1993–95) (1993 election)
17.	Mayawati	June 3, 1995	October 18, 1995	Bahujan Samaj Party	
–	Vacant (President's rule)	October 18, 1995	October 17, 1996	N/A	*Dissolved*
		October 17, 1996	March 21, 1997		Thirteenth Assembly (1996–2002) (1996 election)
(17)	Mayawati Harora	March 21, 1997	September 21, 1997	Bahujan Samaj Party	
(16)	Kalyan Singh Atrauli	September 21, 1997	November 12, 1999	Bharatiya Janata Party	

(continued)

(continued)

S. No.	Name constituency	Term of office		Party	Assembly election
18.	Ram Prakash Gupta	November 12, 1999	October 28, 2000		
19.	Rajnath Singh Haidergarh	October 28, 2000	March 8, 2002		
–	Vacant (President's rule)	March 8, 2002	May 3, 2002	N/A	Fourteenth Assembly (2002–07) (2002 election)
(17)	Mayawati Harora	May 3, 2002	August 29, 2003	Bahujan Samaj Party	
(15)	Mulayam Singh Yadav	August 29, 2003	May 13, 2007	Samajwadi Party	
(17)	Mayawati *None* (MLC)	May 13, 2007	March 7, 2012	Bahujan Samaj Party	Fifteenth Assembly (2007–12) (2007 election)
20.	Akhilesh Yadav *None* (MLC)	March 15, 2012	*Incumbent*	Samajwadi Party	Sixteenth Assembly (2012–17) (2012 election)

Appendix E
Lists of Prime Minister of India

S. No.	Name (Birth–Death) Constituency	Tenure
1.	Jawaharlal Nehru (1889–1964) MP for Phulpur	August 15, 1947–May 27, 1964
Acting	Gulzarilal Nanda (1898–1998) MP for Sabarkantha	May 27, 1964–June 9, 1964
2.	Lal Bahadur Shastri (1904–66) MP for Allahabad	June 9, 1964–January 11, 1966
Acting	Gulzarilal Nanda (1898–1998) MP for Sabarkantha	January 11, 1966–January 24, 1966
3.	Indira Gandhi (1917–84) MP for Rae Bareli	January 24, 1966–March 24, 1977
4.	Morarji Desai (1896–1995) MP for Surat	March 24, 1977–July 28, 1979
5.	Charan Singh (1902–87) MP for Baghpat	July 28, 1979–January 14, 1980
6.	Indira Gandhi (1917–84) MP for Rae Bareli	January 14, 1980–October 31, 1984
	Rajiv Gandhi (1944–91) MP for Amethi	October 31, 1984–December 2, 1989
7.	V.P. Singh (1931–2008) MP for Fatehpur	December 2, 1989–November 10, 1990

(continued)

S. Moinuddin, *Mediascape and The State*, Springer Geography,
DOI 10.1007/978-3-319-51932-6

(continued)

S. No.	Name (Birth–Death) Constituency	Tenure
8.	Chandra Shekhar (1927–2007) MP for Ballia	November 10, 1990–June 21, 1991
9.	P.V. Narasimha Rao (1921–2004) MP for Nandyal	June 21, 1991–May 16, 1996
10.	Atal Bihari Vajpayee (born 1924) MP for Lucknow	May 16, 1996–June 1, 1996
11.	H.D. Deve Gowda (born 1933) MP (Rajya Sabha) for Karnataka	June 1, 1996–April 21, 1997
12.	I.K. Gujral (1919–2012) MP (Rajya Sabha) for Bihar	April 21, 1997–March 19, 1998
13.	Atal Bihari Vajpayee (born 1924) MP for Lucknow	March 19, 1998–May 22, 2004
14.	Manmohan Singh (born 1932) MP (Rajya Sabha) for Assam	May 22, 2004–May 16, 2014
15.	Narendra Modi MP (Lok Sabha) from Varanasi (UP) and Vadodara (Gujarat)	May 26, 2014

Appendix F
Centre–State Relationship Between Uttar Pradesh and Delhi (Centre)

Political parties in power at the state level	Year	Assembly tenure	Political parties in power at the centre level
Congress Party (INC)		1952–67 (Charan Singh of Bhartiya Kisan Party for 10 months during 4th assembly and around 8 months during 5th Assembly session) 1967–89 (Janta Party during 7th Assembly session for period of 3 years)	Congress Party Congress Party
Bhartiya Kranti Dal (Charan Singh)		April 3, 1967 to February 25, 1968	Congress Party
Janta Party (Ram Naresh Yadav and Banarsi Das)	June 23, 1977	February 27, 1979 to June 9, 1980	Janta Party (Morarji Desai and Charan Singh)
Banarsi Das Hapur	February 28, 1979	February 17, 1980	
Janta Dal	1989–90	Mulayam Singh Yadav	National Front (Janta Dal + BJP—V.P Singh was PM)
Janta Dal	June 24, 1991	Tenth assembly (1989–91) (1989 election)	Chandra Shekar (Supported by Congress Party)
	December 6, 1992		
	December 4 1993		

(continued)

S. Moinuddin, *Mediascape and The State*, Springer Geography,
DOI 10.1007/978-3-319-51932-6

(continued)

Political parties in power at the state level	Year	Assembly tenure	Political parties in power at the centre level
BJP	June 3, 1995	Eleventh assembly (1991–92) (1991 election)	Congress Party (Lok Dal of Ajit Singh from UP was major political party who supports the govt.)
N/A SP	October 18, 1995 October 17, 1996	Dissolved Twelfth assembly (1993–95) (1993 election)	Congress Party (P.V. Narasimha Rao) Congress Party
BSP	March 21, 1997		
N/A	September 21, 1997	Dissolved	United Front (SP was supported)
	November 12, 1999	Thirteenth assembly (1996–2002) (1996 election)	NDA (Atal Bihari Vajpayee)
BSP	October 28, 2000		
BJP	March 8, 2002		
N/A	May 3, 2002	Fourteenth assembly (2002–07) (2002 election)	UPA I (Dr. Manmohan Singh)
BSP	August 29, 2003		
SP	May 13, 2007		
BSP	March 7, 2012	Fifteenth assembly (2007–12) (2007 election)	UPA I & II (Dr. Manmohan Singh)
SP	Incumbent	Sixteenth assembly (2012–17) (2012 election)	UPA II (Dr. Manmohan Singh) (16 May 2014)
SP	Incumbent	Do	Narendra Modi (May 26, 2014—onwards)

Appendix G
Popular Slogans Propagated by Political Parties

S. No.	Slogans	Political Party	Year	Region
1.	"UP hui hamari hai, ab dilli ki bari hai" (UP is mine, now will be Delhi); Bharat ki majboori hai, behen Mayawati zaroori hai (Only option in the country, Mayawati is necessary)"	BSP	2000	UP
2.	"Tilak, taraju aur talawar, inko maro joote char (Brahmins, Baniyas and Thakurs should be beaten with shoes)"	BSP	1980–90s	UP
3.	"Hathi nahi Ganesh hain, Brahma, Vishnu, Mahesh hain (the elephant represents Ganesh, and also the trinity, all Brahminical gods, who form the trinity of gods in Hindu mythology)"	BSP	2000	UP
4.	"Aam Aadmi (Common people)"	INC	2000	All India
5.	"Congress ka hath, aam aadmi ke sath (Hand of Congress, along with common people)"	INC	2009	Do
6.	"Aam aadmi ke badhte kadam, har kadam par Bharat buland (Marching steps of Common people, each step has proud of country)"	INC	2000	Do
7.	"Congress ki pechan, vikas aur nirman (Identity of Congress, Development and Construction)"	INC	2000	Do
8.	"Jai Ho (victorious)"	INC	2000	Do
9.	"Majboot neta, nirnayak sarkar (Strong leadership, decisive Government)"	BJP	2000	Do
10.	"Good Governance, Develpoment and Security"	BJP	2000	Do
11.	"India Shining' and 'Feel Good'"	BJP	2000	Do
12.	"Varun-Arun ki kahani, khatam ho gaye Advani"	INC	2000	Do
13.	"Vikas ka mehnatana mangne aaye hain (Demanding pay back for development)"	NDA	2000	Bihar
14.	"Aatank mein jeena hai ya aman-chain mein (Live fearful or live fearless)"?	NDA	2000	Bihar

(continued)

S. Moinuddin, *Mediascape and The State*, Springer Geography,
DOI 10.1007/978-3-319-51932-6

(continued)

S. No.	Slogans	Political Party	Year	Region
15.	"Aapne diya sath to, RSS-VHP-Bajrang Dal par lagega pratibandh (Let mandate will ban RSS-VHP)"	LJP	2000	Bihar
15.	"Varun nahi, yeh aandhi hai, doosra Sanjay Gandhi hai (Varun is a storm like another Sanjay Gandhi)"	BJP	2000	UP
16.	"Mera bharosa aappar, apka viswas mujhpar (My belief on You and Your on me)"	JD-U	2000	Bihar
17.	"Aao milkar saath ladenge, tanashahi, bhrastachar, anyay-o-atyachar nahi sahenge (fight together against corruption, unjustice)"	JD-U	2000	Bihar

Source For more details see TOI on April 4, 2009, http://timesofindia.indiatimes.com/india/Parties-come-out-with-catchy-slogans/articleshow/4391703.cms

Appendix H
Exit Poll Survey of Uttar Pradesh Assembly Election 2012

Exit Poll source	BSP	SP	BJP	INC-RLD	Others
STAR News-Nielsen	101	135	61	99	7
News 24	108	127	57	94	17
India TV-C Voter	122–130	137–145	79–87	39–55	2–10
STAR News-Nielsen	83	183	71	51	11
CNN-IBN	65–70	232–250	28–38	36–44	11–23
Aaj Tak	88–98	195–210	50–56	38–42	20

Source http://electionaffairs.com/uttar%20pradesh/state_assembly_elections_2012/exit_poll_results_uttar_pradesh_Assembly_Elections_2012.html and http://en.wikipedia.org/wiki/Uttar_Pradesh_Legislative_Assembly_election_2012

S. Moinuddin, *Mediascape and The State*, Springer Geography,
DOI 10.1007/978-3-319-51932-6

Appendix I
Summary of the Uttar Pradesh Legislative Assembly Election 2012 Result

Party	Seats contested	Seats won	Seat change	Vote share (%)	Swing (%)
Samajwadi Party		224	▲ 127	29.15	▲ 3.72
Bahujan Samaj Party		80	▼ 126	25.91	▼ 4.52
Bharatiya Janata Party		47	▼ 4	15	▼ 1.97
Indian National Congress		28	▲ 6	11.63	▲ 3.03
Rashtriya Lok Dal		9	▼ 1		
Nationalist Congress Party		1	— 0		
Independents		14			
Total	–	403	–		
Turnout: 59.5%					

Source Election Commission of India 2012

S. Moinuddin, *Mediascape and The State*, Springer Geography,
DOI 10.1007/978-3-319-51932-6

Glossary

Mukhiya	Village head man (It is elected representative)
Chaukidar	Watchman
Zamindar	Zamindar was the land lord during British rule and the Zamindari system was abolished after independent in India
Goonda (ism)	A person have criminal attitude
Chaupal-Chabutra	A common sitting place in the village where villagers sharing their day-to-day activities
Dhaba	A local restaurant generally called dhaba
***घोषणापत्र* (Goshnapatra)**	Election manifesto (Political parties supposed to announce election manifesto prior to election)
***सत्ता* (Satta)**	Governance
***आरक्षण* (Arakshan)**	Reservation
***ऊंची जाति* (Unchi Jatti)**	Forward caste (e.g., Rajput, Brahmin, etc.)
***पुतला फूंका* (Putla Funka)**	Burning Effigy
***मुठभेड़* (Mutbhed)**	Encounter
***कंगाल* (Kangal)**	Bankrupt
बहानेबाजी (Bhanebaji)	Excuse
दौलत की बेटी (Daulat Ki Beti)	Daughter of Rich Person

S. Moinuddin, *Mediascape and The State*, Springer Geography,
DOI 10.1007/978-3-319-51932-6

सौतेला व्यवहार (Sautela Vehawar)	Step Behaviour
विकास (Vikas)	Development
पिछड़ेपन (Pichrepan)	Backwardness
घड़ियाली आंसू (Gharyali Ansuu)	Crocodile tears
सरकारी खजाने (Sarkari Khajane)	Government Treasury
विपक्ष (Wipaksh)	Opposition
विभाजन (Vibhajan)	Division
प्रस्ताव (Prastaw)	Proposal
मूर्तियां' (Murtya)	Statues
मुसलमानों (Musalmano)	Muslims
झूठे वादे (Jhute Wadee)	False Promises
काला झंडा (kala Jhanda)	Black Flag
सांसदों (Sansado)	Parliamentarians
भ्रष्टाचारियों (Bharastcharyeo)	Corrupt peoples
करोड़पति (Crorepati)	Billionaires
नौटंकी (Nautanki)	Play

Bibliography

Adams PC (1992) Television as gathering place. Ann Assoc Am Geogr 82(1):117–135 (Taylor & Francis Ltd)

Adhikaari S (1997) Political geography. Rawat Publications, Jaipur

Adorno TW (2009) Culture industry reconsidered. In: Thornham S et al (eds) Media studies: a reader (3rd ed). Edinburgh University Press, Edinburgh

Aggarwal VB (2002) (ed) Media and society: challenges and opportunities. Concept Publications India, New Delhi

Agnew JA, Duncan JS (2011) Wiley-Blackwell companion to human geography. Wiley-Blackwell, London

Ahmad A (2009) Geography of the South Asian subcontinent: a critical approach. Concept Publications, New Delhi

Ahmed AS (1992) Bombay films: the cinema as metaphor for Indian society and politics. Mod Asian Stud 26(2):289–320

Albarran AB (2013) Social media industries. Routledge, New York

Anderson B, Harrison P (2010) Taking-place: non-representational theories and geography. Ashgate, Surrey

Appadurai A, Carol AB (1995) Public modernity in India. In: Consuming modernity:public culture in a South Asian world (ed), Carol Breckenridge. University of Minnesota Press, Minneapolis

Asp K (1998) Putting hierarchy in its place. Cult Anthropol 3:36–49. Approach. ASHGATE, London

Asp K (1990) Mediatization, Media logic and Mediarchy. Nordicom Rev 11(2):47–50

Barrett M (1991) The politics of truth: from Marx to Foucault. Polity Press, Cambridge

Basu A (2010) Bollywood in the age of new media: the geo-televisual aesthetic. Edinburgh University Press, Edinburgh

Batbyal S (2011) Indian mass media and the politics of change. Routledge, New Delhi

Bathla S (1998) Women, democracy and the media: cultural and political representation in the Indian press. Sage Publications, New Delhi

Baudrillard J (2009) The masses: The implosion of the social in the media. In: Thornham S et al (eds) Media studies: a reader, 3rd edn. Edinburgh University Press, Edinburgh

Bernard B et al (eds) (2010) Simulacra and simulations. In: Greer C (ed) Crime and media: a reader. Routledge, London

Bernard B et al (eds) (2005) Media and mediation. Sage Publications, New Delhi

Bell A (1991) The language of news media. Blackwell, Oxford

Belting H (2011) An anthropology of images (Thomas Dunlop, Trans). Princeton University Press, Princeton

Bennett P (2011) After the media: culture and identity in the 21st century. Routledge, London

Berger J (1971) Ways of seeing. Penguin, London

Bermingham A, Brewer J (eds) (1995) Consumption of culture 1600–1800: image, object text. Routledge, London

S. Moinuddin, *Mediascape and The State*, Springer Geography,
DOI 10.1007/978-3-319-51932-6

Berry C (2003) Mobile cultures: new media in Queer Asia. Duke University Press, Durham
Bhabha HK (ed) (1990) Nation and narration. Routledge, London
Bhabha HK (1994) The location of culture. Routledge, New York
Bourdieu P (1990) The logic of practice. Stanford University Press, Stanford
Bourdieu P (2009). Some properties of field. In: Thornham S et al (eds) Media studies: a reader, 3rd edn. Edinburgh University Press, Edinburgh
Boyd-Barrett et al (eds) (2012) Approaches to Media: A Reader. Rawat Publications, Jaipur
Brass P (1994) The politics of India since independence. Cambridge University Press, Cambridge
Bredl K (2014) Methods for analyzing social media. Routledge, New York
Bucy EP (2002) Living in the information age: a new media reader. Thomson Learning, London
Canty MJ (2007) Image analysis, classification and change detection. CRC Press, Boca Raton
Castells M (1977) The urban question. Edward Arnold, London
Castells M (1983) The city and the grassroots. Edward Arnold, London
Castells M (1996) The rise of network society. Blackwell Publishers, Cambridge
Chattarji S (2008) Tracking the media: interpretations of mass media discourses in India and Pakistan. Routledge, New Delhi
Chaudhary MA, Chaudhary G (2009) Global encyclopaedia of political geography. Global Vision Publications, New Delhi
Chomsky N, Herman ES (2010) A propaganda model. In: Greer Chris (ed) Crime and media: a reader. Routledge, London
Chowdhury A (2011) Electronic media and cultures of communication: a sociological study of the internet (unpublished). University of Delhi, Delhi
Claval P (2012) Region in geography: regional geography: past and present: a review of ideas, approaches and goals. Critical Quest, New Delhi
Clifford N (2010) Key methods in geography. Sage Publications, New Delhi
Cockburn C (1977) The local state. Pluto Press, London
Conboy M (2007) The language of news. Routledge, London
Cook TE (2009) Governing with the news: the news media as a political institution, 2nd edn. University of Chicago Press, Chicago
Corner J, Pels D (2003) Media and the restyling of politics: consumerism, celebrity and cynicism. Sage Publication, London
Cosgrove D (ed) (1999) Mappings. Reaktion, London
Cosgrove D (ed) (2008) Geography and vision: seeing, imaging and representing the world. I. B. TAURIS, New York
Cosgrove D (ed) (2007). Mapping/cartography. In: Atkinson D et al (eds) Cultural geography: a critical dictionary of key concepts. Rawat Publications, Jaipur
Cosgrove D, Stephen D (1988) The iconography of landscape. Cambridge University Press, Cambridge
Cox KR et al (eds) (2008) Sage handbook of political geography. Sage Publications, London
Crang M, Thrift N (2000) Thinking space. Routledge, London
Cresswell T (1996) In place/out of place: geography, ideology, transgression. University of Minnesota Press, Minneapolis
Crush J (ed) (1995) Power of development. Routledge, London
Curtis S (1989) The geography of public welfare provision. Routledge, London
Dahlgren P (1995) Television and the public sphere: citizenship, democracy and the media. Sage Publications, New Delhi
Daniels S (2011) Envisioning landscapes, making worlds: geography and the humanities. Sage Publications, London
Davis H, Walton P (eds) (1983) Language, image and media. Basil Blackwell, Oxford
De Blij H (2012) Why geography matters: more than ever. Oxford University Press, New York
Dear MJ et al (2011) geohuminities: art, history, text at the edge of place. Routledge, New York
Defluer LM, Ball-Rokeach S (1989) Theories of mass communication, 5th edn. Longman, New York

Deleuze G (1986) Cinema 1: the movement-image. Continuum, London
Deleuze G (1989) Cinema 2: the time-image. (trans: Tomlinson H). Continuum, London
Delyser D (2010) Sage handbook of qualitative geography. Sage Publications, London
Devadas NK (2008) Fundamentals of geography. Mohit Publications, New Delhi
Dickey S (1993) Cinema and the urban poor in South India. Cambridge University Press, Cambridge
Dill KE (2009) How fantasy becomes reality: seeing through media influence. Oxford University Press, Oxford
Dirks NB (2001) Castes of mind: colonialism and the making of modern India. Princeton University Press, Princeton
Dobres MA (2000) Technology and social agency. Blackwell, Oxford
Dornfeld B (1998) Producing public television, producing public culture. Princeton University Press, Princeton
Doyle G (2002) Understanding media economics. Sage Publications, London
Dryzek JS (1997) The politics of the earth. Oxford University Press, Oxford
Du Gay P, Pryke M (eds) (2001) Cultural economy. Sage Publications, London
Duncan F (cd) (1996) Bodyspace. Routledge, London
Duncan S, Mark G (1988) The local state and uneven development. Polity Press, Cambridge
Durant A (2010) Meaning in the media: discourse, controversy and debate. Cambridge University Press, New York
Dutt AK, Thakur B (eds) (2012) Facets of social geography: international and Indian perspectives. Cambridge University Press, New Delhi
Eagleton T (2000) The idea of culture. Blackwell, Oxford
ECK DL (1998) Darsan: seeing the divine image in India. Motilal Banarsidass Publishers, New Delhi
Edwards E, Hart J (eds) (2012) India: a sacred geography. Harmony Books, New Delhi
Edwards E, Hart J (eds) (2005) Photographs objects histories: on the materiality of image. Routledge, London
Ellis J (2009) Broadcast TV narration. In: Thornham S et al (eds) Media studies: a reader, 3rd ed. Edinburgh University Press, Edinburgh
Engineer AA (1991) Press on Ayodhya 'Kar Seva'. Economic and political weekly, May 18, pp 1263–1266
Fishman M (1980) Manufacturing the news. University of Texas Press, Austin
Foucault M (2009) Truth and power. In: Thornham S et al (eds) Media studies: a reader, 3rd edn. Edinburgh University Press, Edinburgh
French P, Petley J (1980) Power/knowledge. Pantheon, New York
French P, Petley J (2007) Censoring: the moving image. Seagull, London
Galison P (2001) Image and logic: a material culture of microphysics. Transaction Publications, New Brunswick
Gallaher C (2009) Key concepts in political geography. Sage Publications, Los Angeles
Gans HJ (1979, reprint 2004) Deciding what's news. Pantheon, New York
Gerbaudo P (2012) Tweets and the streets: social media and contemporary activism. Pluto Press, London
Ghose B (2005) Doordarshan days. Penguin, New Delhi
Giddens A (1984) The constitution of society. Polity Press, Cambridge
Giddens A (1985) The nation-state and violence. Polity Press, Cambridge
Giddens A (1990) The consequence of modernity. Stanford University Press, Stanford
Glassner MI (1993) Political geography. Wiley, New York
Godlewska A, Neil S (eds) (1994) Geography and empire. Blackwell, Oxford
Golding P, Elliot P (2009) News values and news production. In: Thornham S et al (eds) Media studies: a reader, 3rd edn. Edinburgh University Press, Edinburgh
Gordon C et al (1980) Power/knowledge: selected interviews and other writings 1972–1977. The Harvester Press, New York

Gorman L, Mc Lean D (2003) Media and Society in the twentieth century. Blackwell, Oxford
Gouldner AW (1979) The future of intellectuals and the rise of the new class. Macmillan, London
Greer C (2010) Crime and media: a reader. Routledge, New York
Gregory D (1994) Geographical imaginations. Blackwell, Oxford
Gregory D, Castree N (2012) Human geography. Sage Publications, London
Gupta A (1995) Blurred boundaries: the discourse of corruption, the culture of politics and the imagined state. Am Ethnologists 22(2):375–402
Hajer M (1995) The politics of environments discourse. Blackwell, Oxford
Hall S et al (2009) The social production of news. In: Thornham S et al (eds) Media studies: a reader, 3rd edn. Edinburgh University Press, Edinburgh
Hall S et al (1980) Cultural studies: two paradigms. Media Cult Soc 2:57–72
Hall S et al (1980) Encoding-Decoding. In: Greer Chris (ed) Crime and media: a reader. Routledge, London
Hammond P (2009) Media, war & post modernity. Routledge, London
Harichandan A, Chaudhary MA (2010) Human geography. Global Vision Publications, New Delhi
Harley JB (2001) The new nature of maps: essays in the history of cartography. John Hopkins University Press, Baltimore
Harley JB (2002) Deconstructing the map. In: Dear Micheal, Flusty Steven (eds) The spaces of postmodernity. Blackwell Publishers, Oxford
Harvey D (1989) The condition of post-modernity. Blackwell, Oxford
Harvey D (1996) Justice, nature and the geography of difference. Blackwell, Cambridge
Harvey D (1982) The limit to capital. Blackwell, Oxford
Hasan Z (1998) Quest for power: oppositional movements and post-congress politics in Uttar Pradesh. Oxford University Press, New Delhi
Held D (1995) Political theory and the modern state. Polity Press, Cambridge
Hermes J (2009) Media, meaning and everyday life. In: Thornham S et al (eds) Media studies: a reader, 3rd edn. Edinburgh University Press, Edinburgh
Highmore B (2002) Everyday life and cultural theory. Routledge, London
Himmelweit H et al (1985) How voters decide. Academic Press, London
Hjarvard S (2008) The mediatization of society. a theory of the media agents of social and cultural change. Nordicom Rev 29(2):105–134
Hoggart R (2004) Mass media in a mass society: myth and reality. Continuum, London
Hooson D (ed) (1994) Geography and national identity. Blackwell, Oxford
Hubbard P et al (2008) Key texts in human geography. Sage Publications, Los Angeles
Hubbard P (2007) Space/place. In: Atkinson D (ed) Cultural geography: a critical dictionary of key concept. Rawat Publications, Jaipur
Israel M (1994) Communication and power: propaganda and the press in the Indian Nationalist Struggle, 1920–1947. Cambridge University Press, Cambridge
Istiaque M (2011) Role of Madarsa's in promoting development and education in Mewat districts: a spatio-temporal studies. ICSSR funded research project (unpublished). New Delhi
Jackson P (1989) Maps of meaning. Unwin Hyman, London
Jeffery R (1993) Indian-language newspaper and why they grow. Econ Polit Wkly 27:2004–2011
Jeffery R (2000) India's newspaper revolution: capitalism, politics and Indian language Press 1977–1999. Oxford University Press, New Delhi
Jeffery R (2010) Media and modernity: communications, women and the state in India. Permanent Black, Ranikhet
Johnson I (2011) Spatiality and the social web: resituating authoritative content. In: Dear M et al (eds) GeoHumanities: art, history, text at the edge of place. Routledge, London
Johnson K (2001) Media and social change: the modernising influences of television in rural India. Media Cult Soc 23:147–169
Johnson NC (2013) Wiley-Blackwell companion to cultural geography. Wiley, West Sussex
Joshi P (1996) Ek nayii bhumikaa pratiiiksha mein hai (A new role in awaiting). Vidura 33(4):45–47

Kaimura M (2006) The state in India: past and present. Oxford University Press, New Delhi
Keane J (Indian reprint 2012) Domocracy and media: without foundations. In: Boyd-Barrett O, Newbold C (eds) Approaches to media: a reader. Rawat Publications, Jaipur
Khilnani S (1997) The idea of India. Hamilton, London
Kitchin R et al (2009) Rethinking maps: new frontiers in cartographic theory. Routledge, London
Krotz F (2007) The meta-process of “mediatization” as a conceptual frame. Glob Media Commun 3(3):256–260
Kumar A (2011) The making of a small state: populist social mobilization and the Hindi Press in the Uttarakhand movement. Orient Blackswan, New Delhi
Laughey D (2008) Key themes in media theory. Rawat Publications, Jaipur
Livingstone D (1992) The geographical traditions: episodes in the history of a contested enterprise. Blackwell, Oxford
Lukinbeal C, Craine J (2009) Geographic media literacy: an introduction. GeoJournal 74:175–182
Lundby K (ed) (2009) Digital storytelling, mediatized stories: self-representations in new media. Peter Lung, New York
MacEachren AM (2008 Indian print) How maps work: representation, visualization, and design. Rawat Publications, Jaipur
Mackinder HJ (1904) The geographical pivot of history. Geographical Journal 23:421–437
Madan TN (1987) Secularism in its place. J Asian Stud 46:747–759
Mandal S (2005) Film and fiction: word into image. Rawat Publications, Jaipur
Mankekar P (1999) Screening culture, viewing politics: an ethnography of television, womanhood and nation in postcolonial India. Duke University Press, Durham
Mann M (1988) States, war and capitalism. Blackwell, Oxford
Manuel P (1993) Cassette culture: popular music and technology in North India. University of Chicago Press, Chicago
Martin D (1980) Breaking of the image: the sociology of christian. Blackwell, London
Massey D (1994) Space, place and gender. Polity Press, Cambridge
Massey D (1991) The political place of locality studies, Environ Plan A 23: 267–281
Mazzoleni G (2008) Mediatization of politics. In: Donsbash W (ed) The international encyclopedia of communication, vol VII. Blackwell, Malden, pp 3047–3051
Mazzoleni G, Schultz W (1999) Mediatization of politics: a challenge for democracy? Polit Commun 16:247–261
McLuhan M (2010) The medium is the message. In: Greer Chris (ed) Crime and media: a reader. Routledge, London
Menzies K (1976) Talcott Parsons and the social image of man. Routledge and Kegan Paul, London
Merrin W (2005) Baudrillard and the media: a critical introduction. Polity Press, Cambridge
Meyer B, Moors A (2006) Religion, media and the public sphere. Indiana University Press, Boolmington
Mitchell WJT (1986) Iconology: image, text, ideology. Chicago University Press, Chicago
Mitra A (1993) Television and popular culture in India: a study of Mahabharat. Sage Publications, New Delhi
Monmonier M (1996) How to lie with maps. The University of Chicago Press, Chicago
Mouffe C (2000) The democratic paradox. Verso, London
Nandy A et al (1995) Creating a nationality. Oxford University Press, Delhi
Nandy A (1988) The politics of secularism and the recovery of religious intolerance. Alternatives 8:177–194
Ninan S (1995) Through the magic window. Penguin, New Delhi
Ninan S, Chattarji S (2013) The hoot reader: media practice in twenty-first century India. Oxford University Press, New Delhi
Ninan S, Chattarji S (1986) Changing agrarian relation in UP: a study of North Eastern area. Inter-India Publications, New Delhi

Painter J, Jeffrey A (2009) Political geography: an introduction to space and power. Sage Publications, New Delhi
Palmer J (1998) News Production. In: Briggs Adam, Cobley Paul (eds) The media: an introduction. Longman, Hong Kong
Palmer S (2006) Television disrupted. Focal Press, Oxford
Pandey G (1990) The construction of communalism in colonial North India. Oxford University Press, Delhi
Pandian MSS (1992) The image trap: M.G. Ramachandran in films and politics. Sage Publications, New Delhi
Partha SV (ed) (2005) Media convergence: issues and experiences. ICFAI University, Hyderabad
Phillips R, Johns J (2012) Fieldwork for human geography. Sage Publications, Los Angeles
Potter R (2012) Key concepts in development geography. Sage Publications, New Delhi
Prasad MM (1998) Ideology of the Hindi film: a historical construction. oxford university Press, Delhi
Rajagopal A (1993) The rise of national programming: the case of the Indian television. Media Cult Soc 15(1):91–110
Rajagopal A (1994) Ram Janmabhumi, consumer identity and image-based politics. Econ Polit Wkly 29(27):1658–1668
Ranganathan M, Rodrigues UM (2010) Indian media in globalised world. Sage Publications, New Delhi
Ranjan S (2006) Television News Coverage. In: Sahay Uday (ed) Making news: handbook of the media in contemporary India. Oxford University Press, New Delhi
Rao NB (2012) Poll surveys in media: an Indian perspective. National Book Trust, New Delhi
Rao U (2010) News as culture: journalistic practices and the remaking of Indian leadership traditions. Berghahn Books, New York
Reeves P (1997) Lucknow Politics: 1920–1947. In: Graff Violette (ed) Memories of a city. Oxford University Press, Delhi
Relph E (1976) Place and placelessness. Pion, London
Rhodes R (1997) Understanding governance. Open University Press, Buckingham
Richard D (1997) Lives of Indian images. University of Princeton Press, Princeton
Rose G (2001) Visual methodologies. Sage Publications, London
Rose G (2003) On the need to ask how, exactly, is geography "visual"? Antipode 35(2):212–221
Rothenbuhler EW, Coman M (eds) (2005) Media anthropology. Sage Publications, California
Roy B, Dutta N (2010) Nandigram and battlelines. Sampark, Kolkata
Rudolph LI (1992) The media and cultural politics. Econ Polit Wkly, July 11: 1489–1496
Sack R (1980) Conceptions of space in social thought: a geographic perspective. MacMillan Press Ltd, Hong Kong
Sack R (1981) Homo geographicus. John Hopkins University Press, Baltimore
Saeed S (2013) Screening the public sphere: media and democracy in India. Routledge, New Delhi
Sahay U (2006) News media interface: a user's perspective. In: Sahay Uday (ed) Making news: handbook of the media in contemporary India. Oxford University Press, New Delhi
Said E (1993) Culture and imperialism. Chatto and Windus, London
Sandbach F (1980) Environment, ideology and policy. Blackwell, Oxford
Saraswati R, Lahiri-Dutt K (2011) Doing gender, doing geography: emerging research in India. Routledge, New Delhi
Schudson M (1995) The power of news. Harvard University Press, Cambridge
Sharma SSP, Singh SP (2012) Growing rural-urban disparity in Uttar Pradesh. Serials Publications, New Delhi
Short J (1993) An introduction to political geography. Routledge, London
Silverstone R (2007) Media and morality: on the rise of mediapolis. Polity, Cambridge
Silverstone R (1994) Television and everyday life. Routledge, London
Silvia T (2001) Global news: perspectives on the information age. Blackwell, Iowa

Singh J (2001) Politics of Harit Pradesh: the case of Western UP as a separate state. EPW, August 4: 2961–2967
Smith D (1994) Geography and Social justice. Blackwell, Oxford
Smith N (1984) Uneven development. Blackwell, Oxford
Soja E (1989) Postmodern geographies: the reassertion of space in critical social theory. Verso, London
Soja E (2009) Thirdspace; towards a new consciousness of space and spatiality. In: Ikas, Wegner (eds) Communicating in the third space. Routledge, New York
Stanczak GC (2007) Visual research methods: image, society and representation. Sage Publications, Los Angeles
Stoddart D (1986) On geography and its history. Blackwell, Oxford
Stromback J (2008) Four phases of mediatization: an analysis of the mediatization of politics. Int J press/polit 13(3):228–246
Sundaram R (2010) Pirate modernity: Delhi's media urbanism. Routledge, London
Talbot M (2007) Media discourse: representation and interaction. Edinburgh University Press, Edinburgh
Taylor P (ed) (1993) Political geography of twentieth century. Belhaven Press, London
Thrift N (1996) Spatial formations. Sage Publications, London
Tremayne M (2007) Blogging, citizenship and the future of media. Routledge, New York
Tuan YF (1977) Space and place: the perspective of experience. University of Minnesota Press, Minneapolis
Tuchman G (1978) Making news. Free Press, New York
Turkle S (1995) Life on the screen: identity in the age of internet. Simon & Schuster, New York
Ward SJ (2013) A global media ethics: problems and perspectives. Wiley-Blackwell, London
Warf B, Arias S (eds) (2009) The spatial turn: interdisciplinary perspectives. Routledge, London
Whatmore S (2002) Hybrid geographies. Sage Publications, London
Wilkinson SI (2004) Votes and violence: electoral competition and communal riots in India. Cambridge University Press, Cambridge
Willis KD (2009) Development: critical approaches in human geography. In: Clifford JN et al (eds) Key concepts in geography. Sage Publications, Los Angeles
Wood D (2010) Rethinking the power of maps. The Guilford Press, New York

Consulted News Websites

www.aajtak.com
www.ddnews.com
www.mib.nic.in
www.ndtv.com
www.newswatch.in
www.tamindiaorg
www.thehoot.org
www.trai.govt.in
www.wikipedia.org
www.zeenews.com
www.thehindu.com
www.rashtryasahara.com
www.toi.com
www.cnnibn.com
www.eci.in

Index

S. Moinuddin, *Mediascape and The State*, Springer Geography,
DOI 10.1007/978-3-319-51932-6

MIX
Papier aus verantwortungsvollen Quellen
Paper from responsible sources
FSC® C105338

If you have any concerns about our products,
you can contact us on
ProductSafety@springernature.com

In case Publisher is established outside the EU,
the EU authorized representative is:
Springer Nature Customer Service Center GmbH
Europaplatz 3, 69115 Heidelberg, Germany

Printed by Libri Plureos GmbH
in Hamburg, Germany